THEORETICAL ELASTICITY

HARVARD MONOGRAPHS IN APPLIED SCIENCE

These monographs are devoted primarily to reports of University research in the applied physical sciences, with especial emphasis on topics that involve intellectual borrowing among the academic disciplines.

1. Matrix Analysis of Electric Networks, P. Le Corbeiller
2. Earth Waves, L. Don Leet
3. Theory of Elasticity and Plasticity, H. M. Westergaard
4. Magnetic Cooling, C. G. B. Garrett
5. Electroacoustics, F. V. Hunt

HARVARD MONOGRAPHS IN
APPLIED SCIENCE
NUMBER 6

THEORETICAL ELASTICITY

CARL E. PEARSON

ENGINEERING DIVISION, ARTHUR D. LITTLE, INC.

1959

Cambridge, Massachusetts

HARVARD UNIVERSITY PRESS

PRINTED IN THE UNITED STATES OF AMERICA
LIBRARY OF CONGRESS CATALOG CARD
NUMBER 59-9283

To Elin and Hugo

PREFACE

This book grew out of lectures which the author gave for a number of years to graduate students in applied mechanics at Harvard University. Its purpose is in part to discuss modern methods of elasticity theory in a form which does not require extensive mathematical experience, and in part to provide a compact and convenient summary of such methods.

For reasons of brevity, and because it is assumed that the reader has had some contact with stress analysis, the emphasis is on general results rather than on detailed examples; an exception is made in the first few chapters, where it is desirable that some manipulative experience in the background topics there covered should be made available to the reader.

It is a pleasure for the author to acknowledge his indebtedness to his colleagues and friends, particularly Professors Bernard Budiansky, George Carrier, and Howard Emmons, for a number of useful discussions and comments concerning specific topics in the book.

CARL E. PEARSON

Cambridge, Massachusetts
December 1958

CONTENTS

I.	Review of Vectors; Index Notation	1
II.	Rotation of a Coordinate System; Cartesian Tensors	30
III.	Stress	46
IV.	Deformation	65
V.	Basic Equations of Linear Elasticity	83
VI.	General Solutions	110
VII.	Variational Methods	137
VIII.	Thermoelasticity	155
IX.	Time-Dependent Problems	172
X.	Nonlinear Elasticity	198
	Index	213

FIGURES

I–1. Test for right- or left-handedness 1
I–2. Examples of right- and left-handed systems 2
I–3. Vector components 3
I–4. Vector addition 3
I–5. Commutative property of vector addition 4
I–6. Formation of scalar product 5
I–7. Formation of vector product 6
I–8. Orientation of unit normal vector 25
I–9. Application of divergence theorem to region with holes 26
II–1. Rotated coordinate system 30
II–2. Determination of angle of rotation 35
III–1. Stress on area element 46
III–2. Typical components of stress tensor 48
III–3. Relation between stress vector and stress tensor components 49
III–4. Stress space 59
III–5. Mohr circle diagram 61
III–6. Complete Mohr circle diagram 62
III–7. Dyadic circle 63
III–8. Geometric construction of dyadic circle 64

TABLES

1. Direction cosines 31
2. Direction cosines for Exercise 1 33
3. Parametric representation of direction cosines 35
4. Direction cosines for example 40
5. Relations among elastic constants 85

THEORETICAL ELASTICITY

I

Review of Vectors; Index Notation

I–1. Coordinate Systems

For the present, we will restrict ourselves to Cartesian coordinate systems (Fig. I–1); for future convenience, the three mutually perpen-

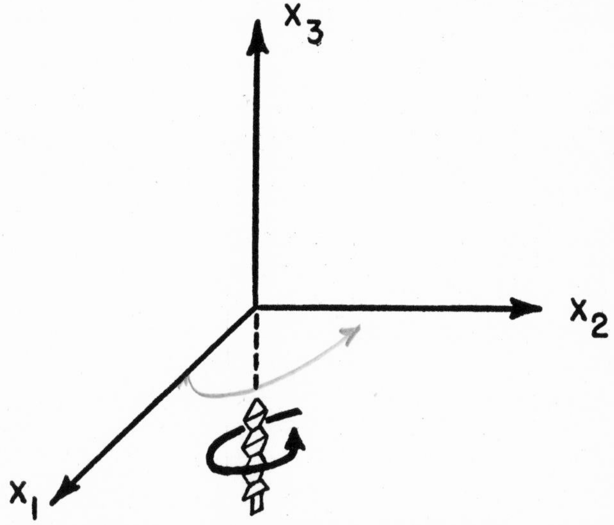

FIG. I–1. Test for right-or left-handedness.

dicular axes will be denoted by x_1, x_2, and x_3 rather than by the more familiar x, y, and z. In the sketch, the x_2 and x_3 axes lie in the plane of the paper and the x_1-axis is directed toward the reader.

Imagine now a right-handed screw whose axis is collinear with the

x_3-axis. If this screw were turned in the direction shown (that is, so as to rotate the x_1-axis into the x_2-axis through the 90° angle — not the 270° angle), it would tend to advance in the direction of the positive x_3-axis. Because of this, the coordinate system is said to be *right-handed.* A system which is not right-handed is called *left-handed.* As examples, A and B in Fig. I–2 are right-handed, but C is left-handed. Note that any

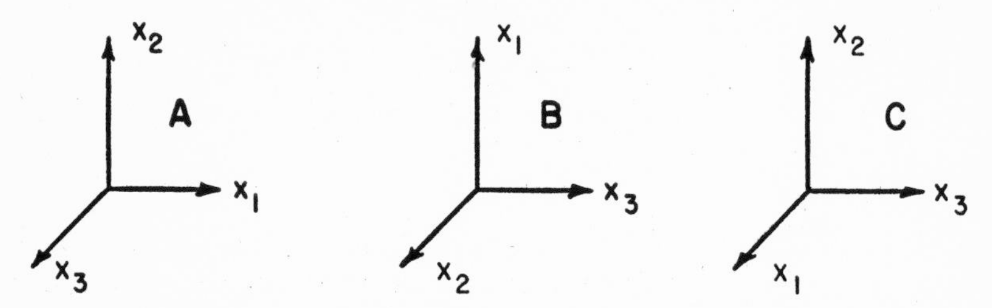

FIG. I–2. Examples of right- and left-handed systems.

two right-handed systems, arranged to have a common origin, may be rotated into one another so that their respective axes coincide. This holds also for any two left-handed systems, but not for one of each. For a reason which will be made clear in the sequel, we will restrict ourselves to the use of right-handed coordinate systems.

I–2. Vector Algebra

A *vector* is a quantity which possesses both magnitude and direction, as contrasted to a *scalar*, which possesses magnitude alone: thus velocity is a vector, temperature a scalar.

A vector is usually represented by an arrow, drawn in the direction of the vector, whose length is made proportional to the magnitude of the vector. A convenient way in which to specify a vector completely is to state the projections of the vector on the three axes of a fixed coordinate system. These projections are called the components of the vector; for a vector **A** we will denote them by A_1, A_2, and A_3 as shown in Fig. I–3. In the sketch, A_2 and A_3 happen to be positive, A_1 negative.

The particular point at which a vector acts is usually understood from the context and need not be specified separately.

Two vectors **A** and **B** are defined to be equal when their respective components are equal; that is, the condition for equality is that

$$A_i = B_i, \quad i = 1, 2, 3.$$

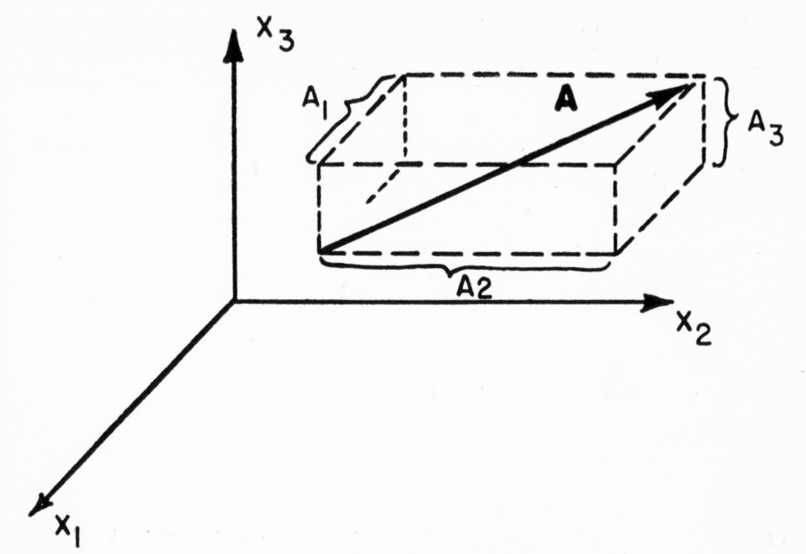

FIG. I–3. Vector components.

In general, equality will be indicated by merely writing

$$A_i = B_i$$

and taking it for granted that, since the subscript i is unspecified, the equation must hold for each of the three possible values of this subscript.

If a vector **A** is multiplied by a positive scalar a, the result $a\mathbf{A}$ is defined (logically enough) to be a new vector coinciding with **A** in direction but of magnitude a times as great. If a is negative, the effect of the negative sign is defined to be a reversal of direction.

The sum **C** of two vectors **A** and **B** is defined according to the parallelogram law as shown in Fig. I–4. Obviously,

$$C_i = A_i + B_i.$$

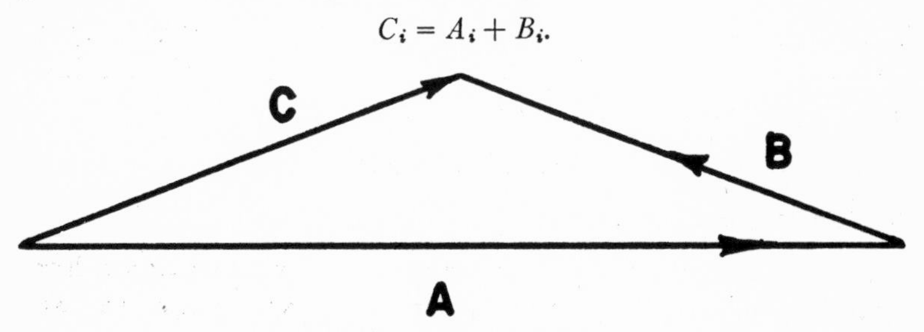

FIG. I–4. Vector addition.

Note that vector addition is commutative, that is (see Fig. I–5),

$$\mathbf{A} + \mathbf{B} = \mathbf{B} + \mathbf{A}.$$

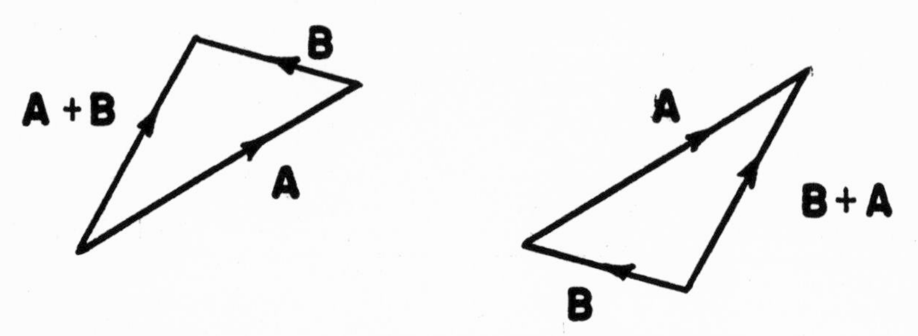

FIG. I–5. Commutative property of vector addition.

It is of course also associative:

$$(\mathbf{A} + \mathbf{B}) + \mathbf{C} = \mathbf{A} + (\mathbf{B} + \mathbf{C}),$$

and so parentheses (indicating which additions are to be performed first) are not required in writing the expression for the sum of a number of vectors.

The magnitude $|\mathbf{A}|$ of a vector $\mathbf{A}$ — proportional to the length of its representative arrow — is defined by

$$|\mathbf{A}| = \sqrt{(A_1^2 + A_2^2 + A_3^2)},$$

as expected. (The square root is always taken as positive.) A vector of unit length is termed a unit vector. Three particularly useful unit vectors are

$$\mathbf{I}, \text{ with components } I_1 = 1,\ I_2 = I_3 = 0,$$
$$\mathbf{J}, \text{ with components } J_2 = 1,\ J_1 = J_3 = 0,$$
$$\mathbf{K}, \text{ with components } K_3 = 1,\ K_1 = K_2 = 0,$$

That is, each is parallel to one of the coordinate axes. In terms of these three unit vectors, any vector $\mathbf{A}$ may be written (using the above definition for the meaning of vector addition):

$$\mathbf{A} = A_1\mathbf{I} + A_2\mathbf{J} + A_3\mathbf{K}.$$

I–3. Scalar Product

We have defined multiplication of a vector by a scalar; we are still free to define multiplication of a vector by another vector in any way we choose. There are two useful ways of combining one vector with another which are akin to ordinary multiplication; these two operations are termed *scalar multiplication* and *vector multiplication*, the names arising because the former operation results in a scalar and the latter in a vector. We will consider only the former in this section.

The *scalar product* (or dot product, or inner product) of **A** and **B** is defined to be the scalar $|\mathbf{A}| \cdot |\mathbf{B}| \cos \theta$, where θ is the angle between **A** and **B** when they are arranged (by parallel translation of one of them, if necessary) so as to have a common initial point, as shown in Fig. I–6.

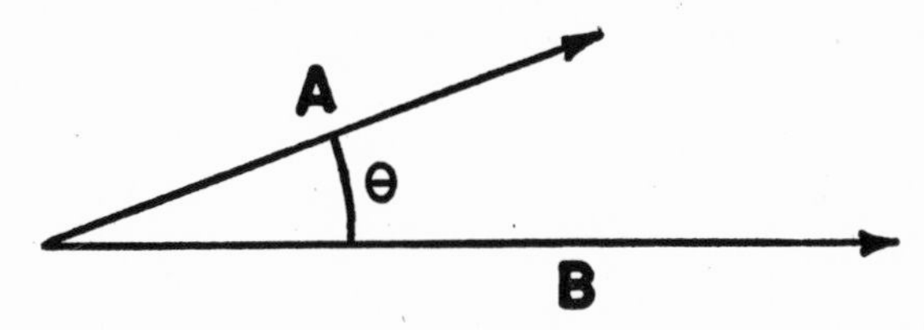

FIG. I–6. Formation of scalar product.

The scalar product is denoted by $\mathbf{A} \cdot \mathbf{B}$. Note that $\mathbf{A} \cdot \mathbf{B}$ may be interpreted geometrically as the product of $|\mathbf{A}|$ by the projection of **B** on **A**, or vice versa.

It follows from the definition that $\mathbf{A} \cdot \mathbf{B} = \mathbf{B} \cdot \mathbf{A}$, and the projection interpretation shows that $\mathbf{A} \cdot (\mathbf{B} + \mathbf{C}) = \mathbf{A} \cdot \mathbf{B} + \mathbf{A} \cdot \mathbf{C}$; hence scalar multiplication is both commutative and distributive. The question of associativity does not enter, since the scalar product of two vectors does not yield another vector and therefore $\mathbf{A} \cdot (\mathbf{B} \cdot \mathbf{C})$ has no meaning; the result of multiplying **A** by the scalar $\mathbf{B} \cdot \mathbf{C}$ would be written $\mathbf{A}(\mathbf{B} \cdot \mathbf{C})$ or $(\mathbf{B} \cdot \mathbf{C})\mathbf{A}$. Note that

$$\begin{aligned} &|\mathbf{A}| = \sqrt{\mathbf{A} \cdot \mathbf{A}}, \\ &\mathbf{I} \cdot \mathbf{I} = \mathbf{J} \cdot \mathbf{J} = \mathbf{K} \cdot \mathbf{K} = 1, \\ &\mathbf{I} \cdot \mathbf{J} = \mathbf{J} \cdot \mathbf{I} = \mathbf{I} \cdot \mathbf{K} = \mathbf{K} \cdot \mathbf{I} = \mathbf{J} \cdot \mathbf{K} = \mathbf{K} \cdot \mathbf{J} = 0. \end{aligned}$$

Using the distributive law, the scalar product of any two vectors may be expressed very simply as

$$\begin{aligned} \mathbf{A} \cdot \mathbf{B} &= (A_1\mathbf{I} + A_2\mathbf{J} + A_3\mathbf{K}) \cdot (B_1\mathbf{I} + B_2\mathbf{J} + B_3\mathbf{K}) \\ &= A_1B_1 + A_2B_2 + A_3B_3 \\ &= \sum_{i=1}^{3} A_iB_i. \end{aligned}$$

We will often encounter sums in which a certain subscript pair is "summed" from 1 to 3, as in the last expression; it will be inconvenient to write summation signs and so we here introduce a summation convention which consists essentially in merely dropping the summation sign.

Summation Convention

Whenever a subscript occurs twice in the same term, it will be understood that the subscript is to be summed from 1 to 3. Thus

$$A_iB_i = A_1B_1 + A_2B_2 + A_3B_3,$$
$$A_pA_p = A_1^2 + A_2^2 + A_3^2.$$

In future $\mathbf{A} \cdot \mathbf{B}$ will be written A_iB_i (or with any other repeated subscript). Such repeated subscripts are often called dummy subscripts because of the fact that the particular letter used in the subscript is not important; thus $A_iB_i = A_sB_s$.

I-4. Vector Product

The result of forming the *vector product* (or cross product) of **A** and **B** is defined to be a new vector **C** whose direction is perpendicular to the plane[1] of **A** and **B** and whose magnitude is given by $\mathbf{C} = |\mathbf{A}| \cdot |\mathbf{B}| \sin\theta$, where θ as before is the angle between **A** and **B**. In Fig. I-7, **A** and **B** lie in the plane of the paper and **C** is perpendicular to it. It is still necessary to specify the orientation of **C** — in this case, whether it is pointing toward or away from the reader. Again a right-handed screw convention is used; if such a screw were placed coaxially with **C** and if **A** (the first-named vector) were to be rotated about **C** into **B** (the second-named vector) through the smaller angle, then **C** is taken to be in the direction in which the screw would tend to move — in this case, toward the reader. The vector product is denoted by

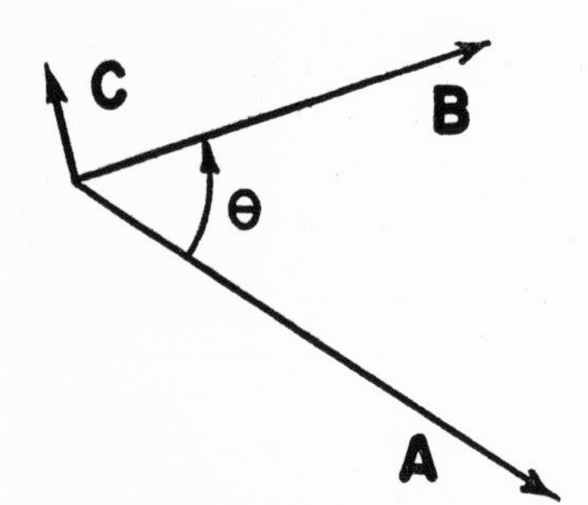

FIG. I-7. Formation of vector product.

$$\mathbf{C} = \mathbf{A} \times \mathbf{B}.$$

[1] Just as in the case of the scalar product, the definition of the vector product does not require that the two vectors have intersecting lines of action. The "plane" of the two vectors as used in the present context means any plane parallel to that plane which contains one vector and is parallel to the second vector.

There are only two cases in which the foregoing rule for finding the orientation of $\mathbf{C}$ breaks down, when $\theta = 0°$ or $180°$, but in these cases $\sin\theta$ and therefore $\mathbf{C}$ are conveniently zero.

Geometrically, the magnitude of $\mathbf{C}$ is seen to be equal to the area of the parallelogram whose sides are $\mathbf{A}$ and $\mathbf{B}$.

It follows from the definition that the vector product is not commutative; in fact, $\mathbf{A} \times \mathbf{B} = -(\mathbf{B} \times \mathbf{A})$. Nor is it associative; the reader may construct an example to show that $\mathbf{A} \times (\mathbf{B} \times \mathbf{C})$ is in general not equal to $(\mathbf{A} \times \mathbf{B}) \times \mathbf{C}$.

The vector product is, however, distributive:

$$\mathbf{A} \times (\mathbf{B} + \mathbf{C} + \mathbf{D} + \cdots) = (\mathbf{A} \times \mathbf{B}) + (\mathbf{A} \times \mathbf{C}) + (\mathbf{A} \times \mathbf{D}) + \cdots.$$

This theorem may be proved by merely noting that, if $\mathbf{X}$ is any vector, then $\mathbf{A} \times \mathbf{X}$ lies in the plane perpendicular to $\mathbf{A}$ and has a length equal to the product of $|\mathbf{A}|$ and the projection of $\mathbf{X}$ onto this plane; simple geometry then yields the result. This distributive property may be used to find a formula for each component of $\mathbf{C} = \mathbf{A} \times \mathbf{B}$ in terms of the components of $\mathbf{A}$ and $\mathbf{B}$. For

$$\mathbf{C} = (A_1\mathbf{I} + A_2\mathbf{J} + A_3\mathbf{K}) \times (B_1\mathbf{I} + B_2\mathbf{J} + B_3\mathbf{K})$$

and if this vector multiplication is carried out term by term then use of such relations as $\mathbf{I} \times \mathbf{J} = \mathbf{K}$, $\mathbf{J} \times \mathbf{I} = -\mathbf{K}$, $\mathbf{K} \times \mathbf{K} = 0$, and so on, yields

$$\mathbf{C} = (A_2B_3 - A_3B_2)\mathbf{I} + (A_3B_1 - A_1B_3)\mathbf{J} + (A_1B_2 - A_2B_1)\mathbf{K},$$

that is, $C_1 = A_2B_3 - A_3B_2$, and so on. An easy way of remembering this formula is to write it in determinantal form, thus:

$$C = \begin{vmatrix} \mathbf{I} & \mathbf{J} & \mathbf{K} \\ A_1 & A_2 & A_3 \\ B_1 & B_2 & B_3 \end{vmatrix}.$$

I–5. Examples

It should be emphasized that the definitions of scalar product and vector product are quite arbitrary; the reader is, in fact, free to introduce new kinds of combination rules for vectors, using any definition he wishes. He will, however, be fortunate if he achieves any combination rules as simple and as useful in applications as the two given above. Perhaps a few examples will be of interest:

(*a*) If a force $\mathbf{F}$ is applied to a body moving with a velocity $\mathbf{V}$, the power input will be $\mathbf{F} \cdot \mathbf{V}$.

(*b*) If a force $\mathbf{F}$ is applied at a point whose radius vector from the origin is denoted by $\mathbf{r}$, then the resultant torques about the three coordinate axes are conveniently represented as the components of the vector $\mathbf{T} = \mathbf{r} \times \mathbf{F}$. (To see this, write $T_1 = r_2F_3 - r_3F_2$ and investigate the physical meaning of each term.)

(*c*) The volume of the parallelepiped constructed upon the three vectors $\mathbf{A}$, $\mathbf{B}$, and $\mathbf{C}$ (having their initial points in common) is $\mathbf{A} \cdot (\mathbf{B} \times \mathbf{C})$.

(*d*) Let it be required to find the shortest distance between two straight wires crossing a stream, the wires not being necessarily in the same plane. Introduce first an arbitrary coordinate system. Let the radius vectors from the origin to the end points of the first wire be $\mathbf{r}_A$ and $\mathbf{r}_B$; for the second wire, let them be $\mathbf{r}_C$ and $\mathbf{r}_D$. Imagine now a straight line drawn between the two wires so as to occupy the path of shortest distance; a vector in the direction of this line is necessarily perpendicular to both of $(\mathbf{r}_A - \mathbf{r}_B)$ and $(\mathbf{r}_C - \mathbf{r}_D)$. Such a vector of unit length is

$$\mathbf{a} = \frac{(\mathbf{r}_A - \mathbf{r}_B) \times (\mathbf{r}_C - \mathbf{r}_D)}{|(\mathbf{r}_A - \mathbf{r}_B) \times (\mathbf{r}_C - \mathbf{r}_D)|}.$$

If now a plane were drawn through each of the wires so as to be perpendicular to $\mathbf{a}$, then the projection on $\mathbf{a}$ of any vector whose initial point lies in one plane and whose terminus lies in the other would equal the perpendicular distance between the planes; such a vector is $(\mathbf{r}_A - \mathbf{r}_C)$ and the solution to the problem is therefore

$$(\mathbf{r}_A - \mathbf{r}_C) \cdot \mathbf{a}.$$

Exercise

Four towers stand at the corners of a 100-ft square. Their heights, proceeding clockwise, are 100 ft, 75 ft, 90 ft, and 55 ft. If two wires are strung diagonally across without sag, show that their distance of closest approach to one another is $600/\sqrt{410}$ ft.

(*e*) Consider a point which moves with a velocity $\mathbf{v}$ in space; let its radius vector from the origin of a fixed coordinate system be denoted by $\mathbf{r}$. In a time interval Δt, the radius vector will alter by an amount $\Delta\mathbf{r}$ (so that the vector addition of $\mathbf{r}$ and $\Delta\mathbf{r}$ gives the new radius vector). Then in analogy with the definition of the derivative of a scalar, we define $d\mathbf{r}/dt$ as the limit of the ratio of $\Delta\mathbf{r}/\Delta t$ as $\Delta t \to 0$. Clearly $d\mathbf{r}/dt = \mathbf{v}$. Because this definition is formally identical with that used for scalars, the usual laws of differentiation must hold. Thus for example

$$\frac{d}{dt}\left(\mathbf{r} \times \frac{d\mathbf{r}}{dt}\right) = \left(\frac{d\mathbf{r}}{dt} \times \frac{d\mathbf{r}}{dt}\right) + \left(\mathbf{r} \times \frac{d^2\mathbf{r}}{dt^2}\right).$$

Here the first vector product vanishes (for the two vectors are collinear); if the acceleration $d^2\mathbf{r}/dt^2$ is directed toward the origin (as in the case of central-force motion) then this acceleration vector is collinear with $\mathbf{r}$ so that the second vector product also vanishes. Recognizing that the magnitude of $\mathbf{r} \times d\mathbf{r}/dt$ is proportional to the rate at which the radius vector sweeps out area, we conclude that in central-force problems this rate must be constant. In the case of planetary motions, this result was discovered empirically by Kepler.

I–6. The Symbol e_{ijk}

This symbol, introduced to obtain convenient representation of the vector product, is defined to be equal to $+1$, -1, or 0, depending on the values of the subscripts.

(*a*) The symbol e_{ijk} is defined to be equal to 0 unless each of the numbers 1, 2, 3 occurs as a subscript. Thus, e_{112}, e_{341}, e_{111}, e_{507} are all zero, but e_{321}, e_{132} are not.

(*b*) If the subscripts are some combination of the numbers 1, 2, 3, then the value of e_{ijk} is $+1$ or -1 depending on whether the order of the subscripts is cyclic or not. Thus $e_{123} = e_{231} = e_{312} = +1$, and $e_{132} = e_{213} = e_{321} = -1$.

Consider now the cross product of $\mathbf{A}$ (whose ith component is A_i) and $\mathbf{B}$ (whose ith component is B_i). If $\mathbf{C} = \mathbf{A} \times \mathbf{B}$, then the ith component of $\mathbf{C}$ is given by

$$C_i = e_{ijk} A_j B_k,$$

where the fact that both subscripts j and k are repeated means that each of these indices must be summed from 1 to 3. As an example,

$$\begin{aligned} C_3 &= e_{3jk} A_j B_k \\ &= e_{311} A_1 B_1 + e_{312} A_1 B_2 + e_{313} A_1 B_3 \\ &\quad + e_{321} A_2 B_1 + e_{322} A_2 B_2 + e_{323} A_2 B_3 \\ &\quad + e_{331} A_3 B_1 + e_{332} A_3 B_2 + e_{333} A_3 B_3 \\ &= A_1 B_2 - A_2 B_1, \end{aligned}$$

since the coefficients of all other terms vanish, and $e_{312} = +1$, $e_{321} = -1$.

It was of course not necessary to write out the product in such detail. In the expansion of the product $e_{3jk} A_j B_k$, every possible combination of the subscripts j and k must occur exactly once, but the e_{ijk}-symbol will vanish for all terms except those where j and k are some combination of 1 and 2 — and there are only two such combinations. As further examples, the reader may check that $C_1 = A_2 B_3 - A_3 B_2$ and $C_2 = A_3 B_1 - A_1 B_3$.

I-7. The Symbol δ_{ij}

The symbol δ_{ij} is defined as being equal to $+1$ if i and j are the same numbers, and 0 otherwise. Thus

$$\delta_{11} = \delta_{22} = \delta_{33} = 1$$
$$\delta_{12} = \delta_{21} = \delta_{13} = \delta_{31} = \delta_{23} = \delta_{32} = 0.$$

Note that, because of the summation convention,

$$\delta_{jj} = \delta_{11} + \delta_{22} + \delta_{33} = 3.$$

As an example of the meaning of the δ_{ij}-symbol, suppose we are told that two vectors, (A_i) and (B_i), are related by the equation

$$A_i = \delta_{ij} B_j.$$

Because j is repeated, there is an imaginary summation sign $\sum_{j=1}^{3}$ in front of the right-hand side of this equation; thus A_i is the sum of three quantities. Now for any particular choice of i (since i is not specified, the equation must hold for all choices of i) — say $i = 2$ — only one of these three terms will possess a nonvanishing δ_{ij}, namely, the one in which j is also 2. Thus

$$\begin{aligned} A_2 &= \delta_{2j} B_j \\ &= \delta_{21} B_1 + \delta_{22} B_2 + \delta_{23} B_3 \\ &= 0 + B_2 + 0, \end{aligned}$$

and, in general,

$$A_i = B_i.$$

It appears therefore that the application of δ_{ij} to B_j has merely substituted i for j in B_j; the δ_{ij}-symbol is therefore often called a substitution operator.

I-8. The e–δ Identity

By actual trial, it is easily checked that

$$e_{ijk} e_{ist} = \delta_{js} \delta_{kt} - \delta_{jt} \delta_{ks}$$

(note that i is summed on the left-hand side). This is the only formula in vector algebra which requires memorization; as an aid to memory, note the symmetric arrangement of the subscripts.

As an example, let $j = 1$, $s = 2$, $k = 2$, $t = 3$. Then the left-hand side is $e_{i12}e_{i23} = e_{112}e_{123} + e_{212}e_{223} + e_{312}e_{323} = 0$ since each term vanishes, and the right-hand side is $\delta_{12}\delta_{23} - \delta_{13}\delta_{22}$ which is also zero.

Let us illustrate a powerful advantage of this e–δ notation. In conventional treatments of vector algebra, it is necessary to memorize (or continually consult tables for) many unwieldy formulas, a typical example being

$$\mathbf{D} = \mathbf{A} \times (\mathbf{B} \times \mathbf{C}) = \mathbf{B}(\mathbf{A} \cdot \mathbf{C}) - \mathbf{C}(\mathbf{A} \cdot \mathbf{B}).$$

Further, considerable algebraic manipulation is required for the proofs of such formulas. In the present notation, all vector identities are automatic. For example, the foregoing identity follows from

$$D_i = e_{ijk}A_j(e_{kst}B_sC_t),$$

where the quantity in parentheses is the kth component of the vector $\mathbf{B} \times \mathbf{C}$; then, since $e_{ijk} = e_{kij}$,

$$\begin{aligned} D_i &= e_{kij}e_{kst}A_jB_sC_t \\ &= (\delta_{is}\delta_{jt} - \delta_{it}\delta_{js})A_jB_sC_t \\ &= B_iA_jC_j - C_iA_sB_s, \end{aligned}$$

which proves the result.

Exercises

1. Show that

(*a*) $\delta_{ij}\delta_{ij} = 3,$
(*b*) $e_{ijk}e_{kji} = -6,$
(*c*) $e_{kks} = 0,$
(*d*) $\delta_{ij}\delta_{jk} = \delta_{ik},$
(*e*) $e_{ijk}A_jA_k = 0.$

2. If $B_i = A_i/(\sqrt{A_kA_k})$, show that B_i is a unit vector.

3. Use the e–δ identity to show that

$$e_{psr}e_{qst} = \delta_{pq}\delta_{rt} - \delta_{pt}\delta_{qr}.$$

4. Use the index notation to prove that

$$\mathbf{A} \cdot (\mathbf{B} \times \mathbf{C}) = (\mathbf{A} \times \mathbf{B}) \cdot \mathbf{C}.$$

5. Use the definition of a determinant to show that

$$\begin{vmatrix} a_{11} & a_{12} & a_{13} \\ a_{21} & a_{22} & a_{23} \\ a_{31} & a_{32} & a_{33} \end{vmatrix} = e_{ijk}a_{1i}a_{2j}a_{3k}.$$

6. If (A_i), (B_i), (C_i) are three noncoplanar vectors and (D_i) an arbitrary vector, show that $D_i = aA_i + bB_i + cC_i$ and find simple formulas for the coefficients (for example, multiply through by $e_{ijk}B_jC_k$).

I–9. Differentiation

Suppose that a scalar ϕ (perhaps the temperature) is defined over a region of space. Then it is possible to form the derivative of ϕ with respect to each of the three coordinates (x_i).

Let

$$G_i = \frac{\partial \phi}{\partial x_i}.$$

Then the three (G_i) can be thought of as the components of a vector $\mathbf{G}$, called the *gradient* of ϕ; the relation is conventionally represented by

$$\mathbf{G} = \nabla \phi,$$

where the symbol ∇ represents a vector operator with the components $\partial/\partial x_1$, $\partial/\partial x_2$, $\partial/\partial x_3$.

In subscript notation, we will use a comma to indicate differentiation; thus,

$$G_i = \frac{\partial \phi}{\partial x_i} = \phi_{,i}.$$

The reason for this notation is that it allows us to indicate the vector character of $(\phi_{,i})$ by using a subscript, just as for other vectors. The fact that this vector has the special property of having its ith component given by the partial derivative with respect to x_i of a scalar ϕ is then indicated by the presence of the comma.

Consider now a vector (A_i) — such as, perhaps, the velocity of a fluid — whose components vary from point to point. Then the derivative of the ith component of $\mathbf{A}$ with respect to x_j would be represented by $A_{i,j}$. A quantity which often occurs in practice is

$$A_{i,i} = A_{1,1} + A_{2,2} + A_{3,3}.$$

It is termed the *divergence* of $\mathbf{A}$. Using the ∇ symbol, it is conventionally written $\nabla \cdot \mathbf{A}$, since

$$\nabla \cdot \mathbf{A} = \left(\mathbf{I}\frac{\partial}{\partial x_1} + \mathbf{J}\frac{\partial}{\partial x_2} + \mathbf{K}\frac{\partial}{\partial x_3}\right) \cdot (\mathbf{I}A_1 + \mathbf{J}A_2 + \mathbf{K}A_3)$$

$$= \frac{\partial A_1}{\partial x_1} + \frac{\partial A_2}{\partial x_2} + \frac{\partial A_3}{\partial x_3},$$

in analogy with the rule for finding the dot product of two vectors. Note that $A_{i,i}$ is a scalar; at any space point, it has only one value, and does not possess the three components that a vector would.

The divergence of $\mathbf{A}$ was obtained by forming the scalar product of ∇ and $\mathbf{A}$; similarly, the cross product of ∇ and $\mathbf{A}$ may be written formally as $\nabla \times \mathbf{A}$, called the *curl* of A; its ith component would be

$$e_{ijk} \frac{\partial}{\partial x_j} A_k = e_{ijk} A_{k,j}.$$

(Note that the ith component of $\nabla \times \mathbf{A}$ is $e_{ijk}A_{k,j}$, not $e_{ijk}A_{j,k}$. The latter is the negative of the ith component of the curl.)

Example

Since $\nabla \times \mathbf{A}$ is a vector, we can find its curl, $\nabla \times (\nabla \times \mathbf{A})$. Let us simplify this expression by use of the e–δ identity. The ith component of the sum is

$$e_{ijk}(e_{kst}A_{t,s})_{,j}$$

where the quantity in parentheses is the kth component of $\nabla \times \mathbf{A}$. Note that k, j, s, t are all summed; i is the only "free" index. Since e_{kst} is a constant, it may be taken outside the parentheses to give

$$e_{ijk}e_{kst}A_{t,sj}.$$

Writing $e_{ijk} = e_{kij}$ and using the e–δ identity yields the final result for the ith component of the curl of $\mathbf{A}$ as

$$A_{j,ji} - A_{i,ss} = \frac{\partial^2 A_j}{\partial x_j \partial x_i} - \frac{\partial^2 A_i}{\partial x_s \partial x_s}$$

(both j and s are summed, of course), which in conventional form would be written

$$\nabla(\nabla \cdot \mathbf{A}) - (\nabla \cdot \nabla)\mathbf{A}.$$

Finally, consider again the vector $(\phi_{,i})$. Its divergence would be written $\phi_{,ii} = \phi_{,11} + \phi_{,22} + \phi_{,33}$ and is called the Laplacian of ϕ; it is often denoted by $\nabla \cdot \nabla\phi$ or $\nabla^2\phi$.

Exercises

1. Using index notation, show that

(*a*) $\nabla \times (\nabla\phi) = 0$,
(*b*) $\nabla \cdot (\nabla \times \mathbf{A}) = 0$.

2. If $\xi_r = \frac{1}{2}e_{rst}V_{t,s}$, show that

(*a*) $\xi_{r,r} = 0$,
(*b*) $V_{i,k} - V_{k,i} = 2e_{ijk}\xi_j$,
(*c*) $4\xi_j\xi_j = V_{j,k}(V_{j,k} - V_{k,j})$.

3. In a certain region of space, the velocity of an air stream has the three components $V_1 = 2x_1 \sin x_2 + \ln x_3$ ft/sec, $V_2 = x_1^2 \cos x_2$ ft/sec, $V_3 = x_1/x_3$ ft/sec. Show that

(*a*) the curl of **V** vanishes,
(*b*) $\mathbf{V} = \nabla\phi$, where $\phi = x_1^2 \sin x_2 + x_1 \ln x_3$,
(*c*) $\nabla \cdot \mathbf{V} = 2 \sin x_2 - x_1^2 \sin x_2 - x_1/x_3^2$.

I–10. Determinants

A further example of the use of the e–δ notation is to be found in determinant theory. In Ex. 5 of Sec. I–8, it is required to show that the value of the determinant whose element in the ith row and jth column is a_{ij} is given by

$$a = e_{ijk}a_{1i}a_{2j}a_{3k}.$$

This result[2] is in fact identical with the conventional definition of a determinant; a slightly more general statement is easily seen to be

$$ae_{stp} = e_{ijk}a_{si}a_{tj}a_{pk}.$$

If each side of this equation is multiplied by e_{mtp} (the repeated indices t and p implying summation, of course), then using the e–δ identity to evaluate $e_{stp}e_{mtp}$ as $2\delta_{ms}$ gives

$$a\delta_{ms} = (\tfrac{1}{2}e_{mtp}e_{ijk}a_{tj}a_{pk})a_{si}.$$

The quantity within parentheses is denoted by A_{mi} and called the *cofactor* of a_{mi}. Thus the equation reads

$$a\delta_{ms} = A_{mi}a_{si}.$$

Using the foregoing definition of the cofactor, it also follows at once that

$$a\delta_{ms} = A_{im}a_{is}.$$

These two expressions give the well-known rule for the expansion of a determinant by cofactors.

Consider next an application to the solution of the linear equation set

$$a_{ij}x_j = b_i,$$

where it is assumed that $a \neq 0$. Multiply both sides by A_{im} to give

$$(A_{im}a_{ij})x_j = b_iA_{im}$$

or

$$a\delta_{mj}x_j = b_iA_{im},$$

that is,

$$x_m = \frac{1}{a}\,b_iA_{im},$$

which is Cramer's rule.

[2] The reader should show that the formula $a = e_{ijk}a_{i1}a_{j2}a_{k3}$ is equivalent.

If the value of the determinant (b_{ij}) is b, then the product ab is given by

$$\begin{aligned} ab &= (be_{ijk})a_{1i}a_{2j}a_{3k} \\ &= (e_{stp}b_{is}b_{jt}b_{kp})a_{1i}a_{2j}a_{3k} \\ &= e_{stp}(a_{1i}b_{is})(a_{2j}b_{jt})(a_{3k}b_{kp}), \end{aligned}$$

which is a determinant whose elements consist of sums of products of appropriate elements of a and b; this is a proof of the familiar rule for multiplication of determinants.

Exercise

Use index notation to show that the value of a determinant is unaltered if an arbitrary multiple of one row is added to any other row.

Consider next the functional relation $z_i = f_i(y_1, y_2, y_3)$, which may be thought of as a transformation from one curvilinear coordinate system to another. The *functional determinant* or *Jacobian* is defined to be the determinant J given by

$$J = e_{ijk} \frac{\partial z_i}{\partial y_1} \frac{\partial z_j}{\partial y_2} \frac{\partial z_k}{\partial y_3}.$$

If conversely the y_i are expressed in terms of the z_i, the functional determinant involving $\partial y/\partial z$ is similarly denoted by J^{-1}. This notation is appropriate, for

$$\begin{aligned} JJ^{-1} &= (Je_{ijk}) \frac{\partial y_i}{\partial z_1} \frac{\partial y_j}{\partial z_2} \frac{\partial y_k}{\partial z_3} \\ &= \left(e_{stp} \frac{\partial z_s}{\partial y_i} \frac{\partial z_t}{\partial y_j} \frac{\partial z_p}{\partial y_k}\right) \frac{\partial y_i}{\partial z_1} \frac{\partial y_j}{\partial z_2} \frac{\partial y_k}{\partial z_3} \\ &= e_{stp} \left(\frac{\partial z_s}{\partial y_i} \frac{\partial y_i}{\partial z_1}\right) \left(\frac{\partial z_t}{\partial y_j} \frac{\partial y_j}{\partial z_2}\right) \left(\frac{\partial z_p}{\partial y_k} \frac{\partial y_k}{\partial z_3}\right) \\ &= e_{stp}\delta_{s1}\delta_{t2}\delta_{p3} = 1. \end{aligned}$$

If the value of $\partial y_i/\partial z_j$ is desired in terms of the various $\partial z/\partial y$, then the set of linear equations

$$\frac{\partial z_r}{\partial y_t} \frac{\partial y_t}{\partial z_j} = \delta_{rj} \quad (j \text{ fixed; } r = 1, 2, 3)$$

could be solved for $\partial y_i/\partial z_j$ just as in a previous paragraph of this section.

Alternatively, both sides of

$$Je_{stp} = e_{rmn} \frac{\partial z_r}{\partial y_s} \frac{\partial z_m}{\partial y_t} \frac{\partial z_n}{\partial y_p}$$

may be multiplied by $e_{sti}\, \partial y_p/\partial z_j$ to give

$$2J \frac{\partial y_i}{\partial z_j} = e_{sti} e_{rmj} \frac{\partial z_r}{\partial y_s} \frac{\partial z_m}{\partial y_t}.$$

It is usual to restrict attention to functional transformations for which $J \neq 0$, so that this equation is then solvable for $\partial y_i/\partial z_j$.

Exercises

1. Multiply both sides of the last equation by e_{pqi} to give

$$J \frac{\partial y_i}{\partial z_j} e_{pqi} = e_{rmj} \frac{\partial z_r}{\partial y_p} \frac{\partial z_m}{\partial y_q}.$$

2. Show that

$$\frac{\partial J}{\partial z_j} = -J \frac{\partial}{\partial y_k} \frac{\partial y_k}{\partial z_j}.$$

I–11. Differential Geometry of Space Curves

Consider the space curve given in parametric form by $x_i = x_i(s)$; for convenience, the arc distance s along the curve is chosen as parameter. Then the vector

$$\alpha_i = \frac{dx_i}{ds}$$

is easily shown to have unit magnitude, and is referred to as the *unit tangent vector*. Differentiating the equation

$$\frac{dx_i}{ds} \frac{dx_i}{ds} = 1$$

gives

$$2 \frac{dx_i}{ds} \frac{d^2 x_i}{ds^2} = 0$$

so that the vector d^2x_i/ds^2 is orthogonal to dx_i/ds. Normalizing this vector by dividing by its magnitude gives a unit vector β_i which is called the *principal normal vector:*

$$\beta_i = \frac{d^2 x_i}{ds^2} \left(\frac{d^2 x_t}{ds^2} \frac{d^2 x_t}{ds^2} \right)^{-\frac{1}{2}}.$$

The unit vector α_i lying along the space curve and the unit vector β_i perpendicular to the space curve define a plane called the *osculating plane.* Geometric visualization of the meaning of $\beta_i \sim d\alpha_i/ds$ shows that, among all those planes that pass through α_i, the osculating plane is that one which adheres most closely to the space curve. The *binormal vector* γ_i is defined by

$$\gamma_i = e_{ijk}\alpha_j\beta_k,$$

and so is perpendicular to the osculating plane. Thus these three unit vectors form an orthogonal triad of vectors; as the point of interest moves along the curve, this triad of vectors alters orientation accordingly.

The derivatives with respect to s of α_i, β_i, γ_i are of some interest. First, $d\alpha_i/ds$ is a vector in the direction of β_i; denote the factor of proportionality by $1/R$, so that

$$\frac{d\alpha_i}{ds} = \frac{1}{R}\beta_i.$$

This quantity $1/R$ is called the *curvature* of the space curve. (The reader may check that it equals the limit of $\Delta\theta/\Delta s$ as $\Delta s \rightarrow 0$, where $\Delta\theta$ is the angle between neighboring principal normals; hence the name. Moreover, a circle of radius R lying in the osculating plane with its center a distance R along β_i is that circle which adheres most closely to the curve.)

Second, differentiating the equation

$$\alpha_i\gamma_i = 0$$

gives

$$\alpha_i\frac{d\gamma_i}{ds} + \frac{d\alpha_i}{ds}\gamma_i = 0.$$

Replacing $d\alpha_i/ds$ by $(1/R)\beta_i$ and using $\beta_i\gamma_i = 0$ gives

$$\alpha_i\frac{d\gamma_i}{ds} = 0,$$

that is, $d\gamma_i/ds$ is perpendicular to α_i. But $d\gamma_i/ds$ is also perpendicular to γ_i (differentiate $\gamma_i\gamma_i = 1$) and so $d\gamma_i/ds$ is collinear with β_i. Write, in analogy with the equation for $d\alpha_i/ds$,

$$\frac{d\gamma_i}{ds} = \frac{1}{T}\beta_i,$$

and note that T, in contrast to R, may be either positive or negative. The quantity $1/T$, called the *torsion*, measures the rate at which a curve is departing from its osculating plane, that is, the angular rate at which adjacent binormals are separating.

Finally, the third of these *Frenet-Serret* formulas is obtained by differentiating

$$\beta_i = e_{ijk}\gamma_j\alpha_k$$

to give

$$\frac{d\beta_i}{ds} = e_{ijk}\gamma_j\frac{1}{R}\beta_k + e_{ijk}\frac{1}{T}\beta_j\alpha_k$$

$$= -\frac{1}{R}\alpha_i - \frac{1}{T}\gamma_i.$$

Exercise

A canonical representation of a space curve in the neighborhood of a point P may be obtained by choosing a local Cartesian coordinate system which coincides with the unit-vector triad at P. In this (y_i) system, show that (if s is measured from P) the equation of the curve is:

$$y_1 = s \qquad - \frac{1}{6}\frac{1}{R^2}s^3 + \cdots,$$

$$y_2 = \frac{1}{2R}s^2 \qquad + \cdots,$$

$$y_3 = \qquad - \frac{1}{6}\frac{1}{RT}s^3 + \cdots,$$

where R, T are evaluated at P.

I–12. Differential Geometry of Surfaces

Let a surface in space be described parametrically by $x_i = f_i(u, v)$. Those curves on the surface for which one of u or v is constant (called parametric lines) may be thought of as forming a curvilinear coordinate net. An arbitrary curve in the surface may then be specified parametrically by two functions $u(t)$, $v(t)$, where t is some suitable parameter. The arc length for such a curve may be calculated by

$$ds^2 = dx_i\,dx_i = \left(\frac{\partial x_i}{\partial u}du + \frac{\partial x_i}{\partial v}dv\right)\left(\frac{\partial x_i}{\partial u}du + \frac{\partial x_i}{\partial v}dv\right)$$

$$= \left(\frac{\partial x_i}{\partial u}\frac{\partial x_i}{\partial u}\right)(du)^2 + 2\left(\frac{\partial x_i}{\partial u}\frac{\partial x_i}{\partial v}\right)(du)(dv) + \left(\frac{\partial x_i}{\partial v}\frac{\partial x_i}{\partial v}\right)(dv)^2$$

$$= E(du)^2 + 2F(du)(dv) + G(dv)^2.$$

Note that both E and G are positive. Unit vectors along the parametric lines would be

$$\frac{1}{\sqrt{E}}\frac{\partial x_i}{\partial u} \quad \text{and} \quad \frac{1}{\sqrt{G}}\frac{\partial x_i}{\partial v}.$$

The cosine of the angle between these unit parametric vectors is equal to their scalar product:

$$\cos\theta = \frac{1}{\sqrt{EG}}\frac{\partial x_i}{\partial u}\frac{\partial x_i}{\partial v} = \frac{F}{\sqrt{EG}},$$

from which it follows that the parametric net is orthogonal if and only if $F = 0$.

The vector $e_{ijk}(\partial x_j/\partial u)(\partial x_k/\partial v)$ is orthogonal to each of the unit parametric vectors and so is perpendicular to the surface. Denote the magnitude of this vector by D, so that D is the positive square root of

$$\left(e_{ijk}\frac{\partial x_j}{\partial u}\frac{\partial x_k}{\partial v}\right)\left(e_{ist}\frac{\partial x_s}{\partial u}\frac{\partial x_t}{\partial v}\right),$$

which by use of the e–δ equality becomes

$$EG - F^2.$$

Thus a unit vector perpendicular to the surface is

$$\zeta_i = \frac{1}{D}\, e_{ijk}\frac{\partial x_j}{\partial u}\frac{\partial x_k}{\partial v},$$

where

$$D^2 = EG - F^2.$$

Consider now an elemental parallelepiped constructed on the parametric intervals du, dv; the vector

$$dS_i = e_{ijk}\left(\frac{\partial x_j}{\partial u}\,du\right)\left(\frac{\partial x_k}{\partial v}\,dv\right) = D\zeta_i\,du\,dv$$

is perpendicular to this area element and has a magnitude (by the definition of cross product) which is equal to the size of the area element. Consequently, dS_i is a vector whose three components are equal in magnitude to the projections of the area element onto the coordinate planes. The size of the area element itself is

$$dS = D\,du\,dv.$$

Exercise

Using the last formula, show that on the surface of a sphere, where $x_1 = R \sin \phi \cos \theta$, $x_2 = R \sin \phi \sin \theta$, $x_3 = R \cos \phi$, the element of area is $R^2 \sin \phi \, d\phi \, d\theta$.

It is occasionally necessary to determine the effect on an area element — both in magnitude and orientation — of a deformation of the surface. Think of the given surface as being embedded in a large volume of rubber, and allow the rubber to be deformed in such a way as to carry the surface along with it. Let the material particle (the rubber molecule) at (y_i) move to (z_i) during this deformation. If now two infinitesimal vectors $(dy_i^{(1)})$ and $(dy_i^{(2)})$ lying in the original surface determine an original surface element

$$dS_i^0 = e_{ijk} \, dy_j^{(1)} \, dy_k^{(2)},$$

then the final surface element will be

$$dS_i = e_{ijk} \, dz_j^{(1)} \, dz_k^{(2)},$$

where

$$dz_i^{(1)} = \frac{\partial z_i}{\partial y_j} dy_j^{(1)} \quad \text{and} \quad dz_i^{(2)} = \frac{\partial z_i}{\partial y_j} dy_j^{(2)},$$

and the question arises of determining the relation between dS_i^0 and dS_i. Combining the last three equations gives

$$dS_i = e_{ijk} \frac{\partial z_j}{\partial y_s} \frac{\partial z_k}{\partial y_t} dy_s^{(1)} \, dy_t^{(2)}$$

and using the result of an exercise of Sec. I–10 gives

$$dS_i = J \frac{\partial y_p}{\partial z_i} dS_p^0.$$

The magnification ratio dS/dS^0 may be obtained by writing

$$dS_i = dS \zeta_i, \quad dS_i^0 = dS^0 \zeta_i^0.$$

Then

$$\frac{dS}{dS^0} = J \left(\frac{\partial y_p}{\partial z_i} \zeta_i \right) \zeta_p^0,$$

which may be interpreted as J times the projection of a directional derivative vector (the quantity in parentheses) onto the original unit normal vector.

Return now to a consideration of the fixed surface $x_i = x_i(u, v)$. If a point P^0 on the surface has coordinates (x_i^0), then the distance of a neighboring point (x_i) on the surface from the tangent plane at P^0 is $\zeta_i^0(x_i - x_i^0)$, where ζ_i^0 is the unit normal vector at P^0. Using a series expansion for $(x_i - x_i^0)$ and the fact that $\zeta_i^0(\partial x_i/\partial u)^0 = 0$, etc., gives this distance as the absolute value of

$$\frac{1}{2}\left[\frac{\partial^2 x_i}{\partial u^2}\zeta_i(du)^2 + 2\frac{\partial^2 x_i}{\partial u\,\partial v}\zeta_i(du)(dv) + \frac{\partial^2 x_i}{\partial v^2}\zeta_i(dv)^2\right]$$
$$= \tfrac{1}{2}\left[e(du)^2 + 2f(du)(dv) + g(dv)^2\right],$$

where the various coefficients are to be evaluated at P^0, and where higher-order terms are omitted. It is easily checked that

$$e(du)^2 + 2f(du)(dv) + g(dv)^2 = -\,dx_i\,d\zeta_i.$$

In analogy with a previous formula for D^2, write

$$d^2 = eg - f^2,$$

but note now that, in contrast with D^2, d^2 need not be positive.

In certain directions from P^0, the distance between surface and tangent plane may vanish to the second order, so that

$$e(du)^2 + 2f(du)(dv) + g(dv)^2 = 0.$$

If all of e, f, g are zero, then this is the case for all directions and P^0 is said to be a *planar* point of the surface. If P^0 is not a planar point, then there are clearly 2, 1, or 0 such *asymptotic directions* for which the quadratic form vanishes, depending on whether $d^2 < 0$, $d^2 = 0$, or $d^2 > 0$; P^0 is then termed a *hyperbolic*, *parabolic*, or *elliptic* point, respectively. At a hyperbolic point, the surface lies partly on one side of the tangent plane and partly on the other side; it cuts the tangent plane along the two asymptotic directions. At a parabolic point, the surface lies wholly on one side of the tangent plane except for a single asymptotic line of contact. At an elliptic point, the surface lies wholly on one side of the tangent plane and there are no asymptotic directions.

The surface curve $u = u(s)$, $v = v(s)$, where s is distance along the curve, is also a space curve of curvature $1/R$. Now

$$\frac{\beta_i}{R} = \frac{d\alpha_i}{ds}$$

from Sec. I–11, so that (since $\zeta_i \alpha_i = 0$)

$$\frac{\zeta_i \beta_i}{R} = \zeta_i \frac{d\alpha_i}{ds} = -\alpha_i \frac{d\zeta_i}{ds}$$

$$= -\frac{dx_i}{ds}\frac{d\zeta_i}{ds}.$$

Substituting the two quadratic forms for $-dx_i\, d\zeta_i$ and ds^2, and recognizing that $\zeta_i \beta_i$ is the cosine of the angle ϕ between the osculating plane of the curve and the normal plane (that is, the plane perpendicular to the tangent plane) through (α_i), this result becomes

$$\frac{\cos\phi}{R} = \frac{e(du)^2 + 2f(du)(dv) + g(dv)^2}{E(du)^2 + 2F(du)(dv) + G(dv)^2}.$$

Since the right-hand side depends only on P^0 and on the direction (dv/du) of the surface curve, it follows that all surface curves passing through P^0 and having the same osculating plane at P^0 have the same curvature at P^0. The right-hand side is usually denoted by $1/r$, and, since it represents the curvature (except for sign, possibly) of a surface curve lying in the normal plane, it is called the *normal curvature* at P^0 for the direction (dv/du). For an asymptotic direction, $1/r = 0$. Note that $R \leq |r|$.

The reader may now show that there always exist two mutually perpendicular directions (dv/du) where $1/r$ attains its maximum and minimum values; the two directions, called *principal directions*, are solutions of

$$(Ef - eF)(du)^2 + (Eg - Ge)(du)(dv) + (Fg - Gf)(dv)^2 = 0.$$

The corresponding extremes of $(1/r)$ are called the *principal normal curvatures* at P^0, and are denoted by $(1/r_1)$ and $(1/r_2)$. The "averaging" quantities

$$\frac{1}{r_1 r_2} \quad \text{and} \quad \frac{1}{r_1} + \frac{1}{r_2}$$

are called the *Gaussian* (or total) and *mean* curvatures respectively.

Exercises

1. A surface curve whose direction is everywhere a principal direction is called a *line of curvature*. Show that the parametric directions are lines of curvature if and only if $F = f = 0$.
2. Show that

$$\frac{1}{r_1 r_2} = \frac{d^2}{D^2}, \quad \frac{1}{r_1} + \frac{1}{r_2} = \frac{Eg - 2Ff + Ge}{D^2}$$

(and hence that the quantities on the right-hand sides of these two equations are independent of the parametric net used). The easy way of proving these results is to notice that if $(1/r)$ has a maximum or minimum value then the equation relating $(1/r)$ to (du/dv) can have only one solution for (du/dv) so that its discriminant must vanish.

3. Show that if the parametric directions are lines of curvature, with $(dv = 0)$ corresponding to $(1/r_1)$ and $(du = 0)$ to $(1/r_2)$, then

$$\frac{1}{r} = \frac{1}{r_1}\cos^2\theta + \frac{1}{r_2}\sin^2\theta,$$

where $1/r$ is the normal curvature for a curve direction making an angle θ with $(dv = 0)$.

4. Show that the sum of the normal curvatures in two mutually perpendicular directions does not depend upon those directions.

5. Show that the asymptotic directions at a hyperbolic point are equally inclined to each principal direction at the point, and that they are perpendicular to each other if and only if the mean curvature is zero.

I–13. Surface-Volume Integrals. Further Properties of Vectors

A. Gauss's and Stokes's Theorems

In continuum mechanics, the total force on a given volume of matter arises partly from internal body forces (such as gravity) and partly from surface forces (such as pressure or friction); these two types of forces may be written as volume and surface integrals respectively. In order to combine the two so as to calculate the net force, it is necessary to be able to express a surface integral as a volume integral or vice versa. It will turn out that the following *divergence theorem* of Gauss is applicable in such cases:

Given any vector field (B_i) defined throughout a volume V (that is, at every point of V the components B_i are known), then

$$\int B_{i,i}\, dV = \int B_i n_i\, dS,$$

where the left-hand side is the volume integral of the divergence and the right-hand side is the surface integral of the *flux* $B_i n_i$, n_i being the outward unit normal vector at the surface element dS.

A physical interpretation of this theorem is easy, and serves also to indicate the method of proof. Imagine an incompressible fluid moving with vector velocity (B_i) at each point. It may of course be necessary to allow sources and sinks to exist, since (B_i) can easily be such as to result in an accumulation or disappearance of fluid in certain regions. If an infinitesimal parallelepiped (dx_1, dx_2, dx_3) is constructed at a point P, then the net outflow of fluid is easily calculated to be $(B_{i,i})\, dx_1\, dx_2\, dx_3$; filling up V with these parallelepipeds then gives the theorem at once (note that $B_i n_i$ is the velocity component in the direction of the outward normal to dS).

This theorem indicates the reason for the name "divergence" for the quantity $(B_{i,i})$, for this quantity corresponds physically to source strength per unit volume (divergence of fluid). In fact, the divergence of (B_i) is occasionally *defined* as the limit of

$$\frac{1}{V}\int B_i n_i\, dS$$

as $V \to 0$. The theorem then shows that this alternative definition is equivalent to that previously given.

The two-dimensional case where all quantities depend only on x_1 and x_2 results from writing the foregoing theorem for an indefinitely long cylinder with generators parallel to the x_3-axis and passing through a given closed curve C in the x_1x_2-plane:

$$\int (B_{1,1} + B_{2,2})dA = \int (B_1n_1 + B_2n_2)dl,$$

where dl is the element of arc length along C and dA is the element of area in the x_1x_2-plane.

Exercises

1. Using the plane form of the divergence theorem with a new vector (D_i) defined by $D_1 = B_2$, $D_2 = -B_1$, and noting that $n_1\, dl = dx_2$, $n_2\, dl = -dx_1$ (for counterclockwise traversal of boundary), show that

$$\int (D_{2,1} - D_{1,2})dA = \int (D_1\, dx_1 + D_2\, dx_2)$$

and interpret in terms of curl.

2. Note that the theorem still holds if only one component of the given vector is nonzero. For example, set $D_1 = x_2$, $D_2 = 0$ in the previous exercise to obtain

$$A = -\int x_2\, dx_1.$$

3. Setting $B_i = FG_{,i}$, where F and G are arbitrary functions, show that

$$\int FG_{,ii}\, dV + \int F_{,i}G_{,i}\, dV = \int F\frac{\partial G}{\partial n}\, dS,$$

and note the special cases that ensue for $F = 1$, $F = G$, $G_{,ii} = 0$, and so on; note also the possibility of subtracting from this equation the corresponding equation with F and G transposed. Making use of one of these *Green's identities* show, by considering the difference of two possible solutions, that any solution of the problem [$G_{,ii} = 0$, G prescribed on the boundary] must be unique.

4. Show that a physical interpretation of $G_{,ii}$ is given by the statement that it is proportional to the difference between the value of G at P and the mean value $\bar{G}$ of G over the surface of a sphere of radius r surrounding P; in fact, $\frac{1}{6}r^2G_{,ii} = \bar{G} - G$, where higher-order terms are omitted.

An extension of the result of Ex. 1 is associated with the name of Stokes. Consider any spatial surface S with boundary C, immersed in a region throughout which a vector field (B_i) is defined. Then

$$\int e_{ijk}B_{k,j}n_i\, dS = \int B_i\, dx_i,$$

where n_i is the unit normal corresponding to the surface element dS, and (dx_i) is the elemental vector of the contour C. The sense in which the contour must be traversed is similar to that required in Ex. 1 above (see Fig. I–8). The theorem is proved by merely noting that it must be true for an individual surface element (for example, use a "local" coordinate system [3] for each area element and also the result of Ex. 1); adding up over the whole surface then gives the result.

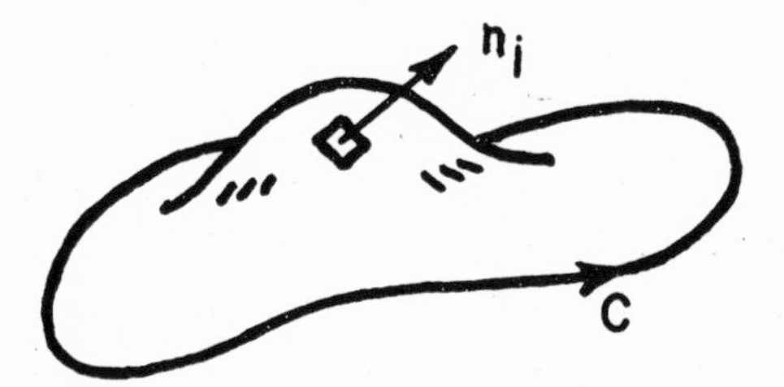

FIG. I–8. Orientation of unit normal vector.

Exercises

1. Construct algebraically a simple vector field B_i and verify by actual calculation the theorems of Gauss and Stokes for some simple volume and surface.
2. Extend the statements of the two theorems to cover regions with "holes"

[3] Strictly speaking, we need the fact that a vector result true in one Cartesian coordinate system is true in all such systems; this will follow from Chapter II.

(Fig. I–9). (Note that unit normals must always be directed away from the region, and so will point into holes.)

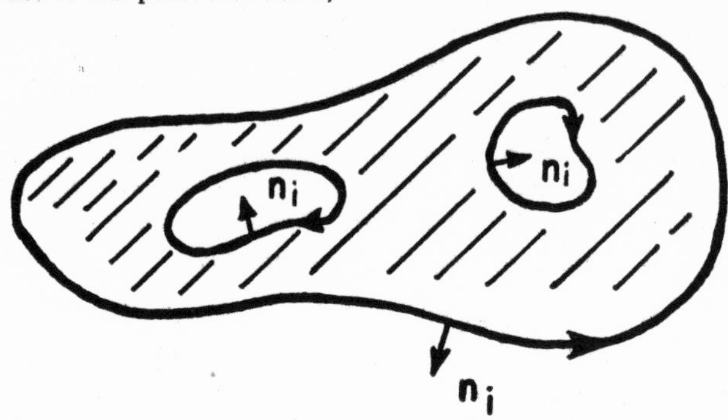

FIG. I–9. Application of divergence theorem to region with holes.

3. In the same way as was done above for divergence, give an alternative definition of curl (note the necessity of considering the orientation of the area element). These alternative definitions are often useful for calculating the expression for divergence and curl in curvilinear coordinate systems.

B. Corollary to Stokes's Theorem

The vector field (B_i) is termed *irrotational* if $e_{ijk}B_{k,j} = 0$ everywhere. (If B_i is a fluid velocity vector, then this condition implies that there is no local fluid rotation, hence the name.) An irrotational vector field has a number of properties:

(*a*) The integral $\int B_i \, dx_i$ around any closed curve must, by Stokes's theorem, be zero;

(*b*) The integral $\int B_i \, dx_i$ between any two points must be independent of the path of integration (otherwise result (*a*) would be violated for a closed contour made up of the two paths);

(*c*) It now follows from (*b*) that the integral

$$\int_{P_0}^{P} B_i \, dx_i,$$

where P_0 is some fixed point, P an arbitrary point, defines a function $\phi(P)$ the value of which does not depend on the path of integration. Moreover, a consideration of infinitesimal motion of P in the direction of any axis shows that

$$\frac{\partial \phi}{\partial x_i} = B_i.$$

This last result is often expressed by saying (in analogy with gravitational fields) that the vector (B_i) is derivable from a scalar potential function ϕ.

It is easy to show that, if (B_i) is such that any one of (*a*), (*b*), (*c*), or $e_{ijk}B_{k,j} = 0$ holds, then so do the remaining three conditions.

Exercises

1. A particular two-dimensional fluid-velocity vector field (D_i) satisfies $D_{1,1} + D_{2,2} = 0$. Show that there exists a *stream function* G such that $D_1 = G_{,2}$, $D_2 = -G_{,1}$. Using the fact that $(G_{,i})$ is perpendicular to the contour curves of G, show that the velocity vector is tangent to these contour curves.
2. In the two-dimensional case, extend the corollary so as to hold for regions with holes by allowing ϕ to be multiple-valued where necessary.

C. *Solenoidal Vectors. Helmholtz's Theorem*

In part B, it was shown that $B_i = \phi_{,i}$ followed from the condition $e_{ijk}B_{k,j} = 0$. It is natural to ask if the *solenoidal* condition $B_{i,i} = 0$ would imply any similar results. A clue is obtained by noticing that, if B_i were the curl of some other vector, then $B_{i,i} = 0$, consequently we are led to consider the converse of this statement, namely, does there exist a vector R_i such that $B_i = e_{ijk}R_{k,j}$? The components (R_i) would have to satisfy

$$\begin{aligned} R_{3,2} - R_{2,3} &= B_1, \\ R_{1,3} - R_{3,1} &= B_2, \\ R_{2,1} - R_{1,2} &= B_3. \end{aligned}$$

Try $R_3 \equiv 0$. Then

$$R_2 = -\int_0^{x_3} B_1\,dx_3 + f_2(x_1, x_2),$$

$$R_1 = \int_0^{x_3} B_2\,dx_3 + f_1(x_1, x_2),$$

where f_1, f_2 are arbitrary functions of x_1 and x_2. (Of course, any convenient choice for lower limit of integration may be made.) The third equation requires (using now the condition $B_{i,i} = 0$).

$$B_3(x_1, x_2, x_3) - B_3(x_1, x_2, 0) + f_{2,1} - f_{1,2} = B_3(x_1, x_2, x_3),$$

that is,

$$f_{2,1} - f_{1,2} = B_3(x_1, x_2, 0),$$

an equation which can be satisfied by many choices of f_1 and f_2. It has now been shown that a vector (R_i) does indeed exist such that

$B_i = e_{ijk}R_{k,j}$, provided only that $B_{i,i} = 0$. The most general such *vector potential* may be found by noting that if (S_i) is any other vector such that $B_i = e_{ijk}S_{k,j}$, then $e_{ijk}(S_k - R_k)_{,j} = 0$, so that by part B

$$S_k = R_k + \psi_{,k},$$

where ψ is an arbitrary scalar function. Consequently, the condition $B_{i,i} = 0$ implies that $B_i = e_{ijk}S_{k,j}$, where the most general (S_k) vector is

$$S_1 = \int_0^{x_3} B_2\, dx_3 + f_1(x_1, x_2) + \psi_{,1},$$

$$S_2 = -\int_0^{x_3} B_1\, dx_3 + f_2(x_1, x_2) + \psi_{,2},$$

$$S_3 = \psi_{,3},$$

where f_1 and f_2 are any functions satisfying

$$f_{2,1} - f_{1,2} = B_3(x_1, x_2, 0).$$

In particular, we could choose

$$f_2 = 0 \quad \text{and} \quad f_1 = -\int_0^{x_2} B_3(x_1, x_2, 0)dx_2.$$

Note that $S_{i,i}$ may be made equal to any desired function (such as zero) by choosing $\psi_{,ii}$ appropriately. (In general, a function ψ can be found so that $\psi_{,ii}$ has any desired functional form — for this is merely Poisson's equation in electrostatics, which must have a solution by physical considerations.)

Consider next an arbitrary vector field (C_i). It was proved by Helmholtz that

$$C_i = \phi_{,i} + e_{ijk}S_{k,j},$$

that is, that any vector field can be decomposed into an irrotational and a solenoidal portion. Now, if such a decomposition is possible, taking the divergence yields

$$\phi_{,ii} = C_{i,i}.$$

Conversely, if ϕ is any such function (one always exists; see the parenthetical remark in last paragraph), then $(C_i - \phi_{,i})$ is solenoidal, so that the result of the last paragraph applies and the theorem is proved. Again, the fact that (S_i) can be found such that $S_{i,i} = 0$ or $S_{i,i} = C_{i,i}$ is often useful.

Exercises

1. Put the foregoing formula for the vector potential of a solenoidal vector into a more symmetric form by rewriting it cyclically (for $R_1 = 0$, $R_2 = 0$ in turn), adding, and dividing by 3.

2. A vector (B_i) is such that $\int B_i C_i \, dV = 0$ for all vectors (C_i) satisfying $C_{i,i} = 0$ in V, $C_i n_i = 0$ on S. Show that B_i must be irrotational. (*Hint:* Choose C_i as the sum of the solenoidal component of B_i and the gradient of an appropriate harmonic function.)

II

Rotation of a Coordinate System; Cartesian Tensors

II–1. Table of Direction Cosines

The study of elasticity requires a knowledge of the relations which exist between two Cartesian coordinate systems one of which is rotated from the other; the purpose of this chapter is to discuss these relations. Suppose that two right-handed Cartesian coordinate systems are available, as in Fig. II–1. A point in space may be specified by giving its

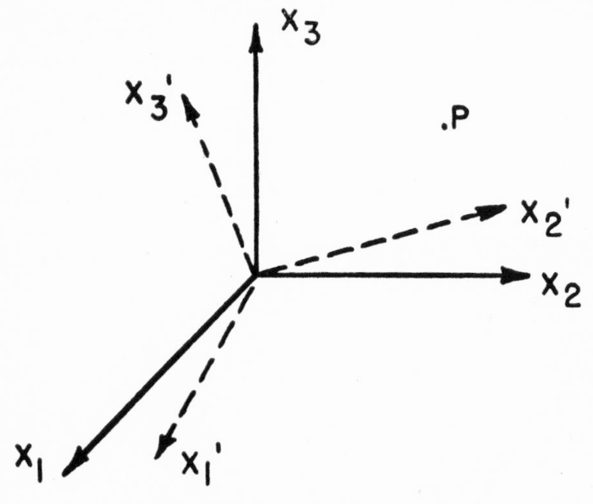

FIG. II–1. Rotated coordinate system.

coordinates in either system, namely, (x_1, x_2, x_3) or (x_1', x_2', x_3'). Let α_{11} be the cosine of the angle between the positive (x_1') and (x_1) axes,

α_{12} the cosine of the angle between the positive (x_1') and (x_2) axes, and so on; these cosines may be conveniently exhibited as in Table 1.

TABLE 1. Direction cosines.

Axis	Axis x_1	x_2	x_3
x_1'	α_{11}	α_{12}	α_{13}
x_2'	α_{21}	α_{22}	α_{23}
x_3'	α_{31}	α_{32}	α_{33}

If the point P (Fig. II–1) has coordinates (x_i) in the unprimed system, and (x_i') in the primed system, then

$$x_i' = \alpha_{ij}x_j \quad \text{and} \quad x_i = \alpha_{ji}x_j',$$

since a unit vector in the direction of the positive (x_2') axis, say, would have components in the (x_i) system $(\alpha_{21}, \alpha_{22}, \alpha_{23})$. The projection of OP onto this unit vector is given by the scalar product

$$x_1\alpha_{21} + x_2\alpha_{22} + x_3\alpha_{23} = \alpha_{2j}x_j,$$

since the components of OP in the (x_i) system are simply (x_1, x_2, x_3). But this projection is the (x_2') component of P in the primed system, and so $x_2' = \alpha_{2j}x_j$. Similarly $x_i' = \alpha_{ij}x_j$ for any choice of i. The proof that $x_i = \alpha_{ji}x_j'$ is similar.

It follows that

$$\alpha_{ij} = \frac{\partial x_i'}{\partial x_j} = \frac{\partial x_j}{\partial x_i'}.$$

II–2. Relations Between the α_{ij}

It will first be shown that

$$\alpha_{ij}\alpha_{kj} = \delta_{ik} \quad \text{and} \quad \alpha_{ji}\alpha_{jk} = \delta_{ik}.$$

The theorem will be proved for only two special cases of the first formula, namely, $i = k = 2$ and $i = 2$, $k = 3$; the general result then follows similarly.

A unit vector in the direction of the positive (x_2') axis has direction cosines $(\alpha_{21}, \alpha_{22}, \alpha_{23})$ with respect to the (x_i) system; since it is a unit vector,

$$\alpha_{21}^2 + \alpha_{22}^2 + \alpha_{23}^2 = 1,$$

which proves the first formula for the case $i = k = 2$.

A unit vector in the direction of the positive (x_3') axis has direction cosines $(\alpha_{31}, \alpha_{32}, \alpha_{33})$ with respect to the (x_i) system. These two unit vectors — in the direction of the (x_2') and (x_3') axes, respectively — are orthogonal, and hence their scalar product

$$\alpha_{21}\alpha_{31} + \alpha_{22}\alpha_{32} + \alpha_{23}\alpha_{33}$$

must vanish; thus the first formula is proved for the case $i = 2$, $k = 3$.

The general case follows similarly. The second formula is proved in the same manner, except that calculations are now carried out in the primed system.

As a second result, the following determinant will be shown to have value unity:

$$D = \begin{vmatrix} \alpha_{11} & \alpha_{12} & \alpha_{13} \\ \alpha_{21} & \alpha_{22} & \alpha_{23} \\ \alpha_{31} & \alpha_{32} & \alpha_{33} \end{vmatrix}.$$

Since interchanging rows and columns does not affect the value of D,

$$D^2 = \begin{vmatrix} \alpha_{11} & \alpha_{12} & \alpha_{13} \\ \alpha_{21} & \alpha_{22} & \alpha_{23} \\ \alpha_{31} & \alpha_{32} & \alpha_{33} \end{vmatrix} \cdot \begin{vmatrix} \alpha_{11} & \alpha_{21} & \alpha_{31} \\ \alpha_{12} & \alpha_{22} & \alpha_{32} \\ \alpha_{13} & \alpha_{23} & \alpha_{33} \end{vmatrix} = \begin{vmatrix} 1 & 0 & 0 \\ 0 & 1 & 0 \\ 0 & 0 & 1 \end{vmatrix} = 1,$$

where the rule for multiplication of determinants has been used and the preceding identity applied. It follows that $D = \pm 1$; but, since D is continuous in the α_{ij} and equals $+1$ for the identity transformation (when the primed and unprimed systems coincide), it must equal $+1$ always.

Third, each element α_{ij} of the determinant D is equal to its own cofactor. For multiplying both sides of

$$\alpha_{ij}\alpha_{kj} = \delta_{ik}$$

by A_{ks} (cf Sec. I–10) gives

$$\alpha_{ij}\delta_{js}D = \delta_{ik}A_{ks},$$

that is, using $D = 1$,

$$\alpha_{is} = A_{is} = \tfrac{1}{2}e_{itp}e_{sjk}\alpha_{tj}\alpha_{pk}.$$

For example,

$$\alpha_{12} = \alpha_{23}\alpha_{31} - \alpha_{33}\alpha_{21},$$

which equation may be used to show that in an infinitesimal rotation of coordinate systems (that is, when the primed and unprimed systems almost coincide), $\alpha_{12} \cong -\alpha_{21}$; similar results hold for the other off-diagonal terms.

Exercises

1. Two Cartesian coordinate systems are related by the cosine table shown in Table 2. Check that each identity of this section holds true. Show also that

TABLE 2. Direction cosines for Exercise 1.

	Axis		
Axis	x_1	x_2	x_3
x_1'	$\frac{12}{25}$	$-\frac{9}{25}$	$\frac{4}{5}$
x_2'	$\frac{3}{5}$	$\frac{4}{5}$	0
x_3'	$-\frac{16}{25}$	$\frac{12}{25}$	$\frac{3}{5}$

the point whose coordinates in the (x_i) system are $(0, 1, -1)$ coincides with the point whose coordinates in the (x_i') system are $(-\frac{29}{25}, \frac{4}{5}, -\frac{3}{25})$. Finally, prove that the following two planes coincide:

$$2x_1 - \tfrac{1}{3}x_2 + x_3 = 1,$$
$$\tfrac{47}{25}x_1' + \tfrac{14}{15}x_2' - \tfrac{21}{25}x_3' = 1.$$

2. Consider a set of nine numbers (β_{ij}) for which $\beta_{ij}\beta_{kj} = \delta_{ik}$ and for which $B = \det. (\beta_{ij}) \neq 0$. Prove that $\beta_{ji}\beta_{jk} = \delta_{ik}$ and $B^2 = 1$ (if $B = +1$, then the the hypothesis of this exercise is the condition that the β_{ij} correspond to a physical rotation of axes).

II–3. Euler's Rotation Formula

Lemma: Any change in position of a rigid body one point of which is held fixed can also be achieved by a single rotation about a definite axis through that point.

For consider a *fixed* coordinate system whose origin coincides with the fixed point of the body. Choose two material particles inside the body, and draw the position vectors from the origin to these two points. In the initial position of the body, denote these vectors by (A_i) and (B_i); the resultant motion of the body is then completely specified by giving their final positions (A_i') and (B_i'). Now, if the final position could be obtained by means of a rotation about an axis through the origin, (A_i) and (A_i') would lie in the surface of a cone encircling this axis and having its vertex at the origin; so would (B_i) and (B_i'). But any cone containing

(A_i) and (A_i') must have its axis in that plane which is the perpendicular bisector of the triangle formed by (A_i) and (A_i'); a similar result holds for (B_i) and (B_i'). But both of these planes can actually be constructed, and since both contain the origin they must intersect along a line. Conversely, this line is clearly the required axis (remember that during the solid-body rotation the two vectors move as a unit).

Euler's formulas

Consider any two Cartesian coordinate systems, (x_i) and (x_i'), having their origins in common. From the lemma just proved, there exists an axis about which either system may be rotated into the other. Let (A_i) be a unit vector in the direction of this rotation axis. Imagine a right-handed screw coaxial with (A_i); if the (x_i) axis system is rotated about the (A_i) axis in such a sense as to make the screw move in the vector direction (A_i), then let the total angle of rotation required to bring the (x_i) system into coincidence with the (x_i') system be denoted by θ.

Let P be any fixed point in space, with coordinates (x_i) and (x_i') in the unprimed and primed systems respectively. If the vector OP were rotated through an angle $-\theta$ about (A_i), both coordinate systems being held fixed, then the new (x_i) coordinates of P would be equal to its previous (x_i') coordinates. Let B, C, D be coplanar vectors of the same length (Fig. II–2), where B, C, A are mutually perpendicular. Then the new (x_i) coordinates of OP — or, equivalently, the desired (x_i') coordinates — are

$$\begin{aligned} x_i' &= D_i + (x_s A_s) A_i \\ &= B_i \cos\theta + C_i \sin\theta + (x_s A_s) A_i. \end{aligned}$$

Using

$$\begin{aligned} B_i &= x_i - A_i(x_s A_s), \\ C_i &= e_{ijk} B_j A_k, \end{aligned}$$

and simplifying gives finally

$$x_i' = x_i - 2\sin^2\frac{\theta}{2}(x_i - x_s A_s A_i) + 2\sin\frac{\theta}{2}\cos\frac{\theta}{2}\, e_{ijk} x_j A_k.$$

As a check, note that if $x_i = \alpha A_i$, then $x_i' = x_i$. The table of direction cosines may now be written down (Table 3), where

$$\xi = A_1 \sin\frac{\theta}{2}, \quad \eta = A_2 \sin\frac{\theta}{2}, \quad \zeta = A_3 \sin\frac{\theta}{2}, \quad \rho = \cos\frac{\theta}{2},$$

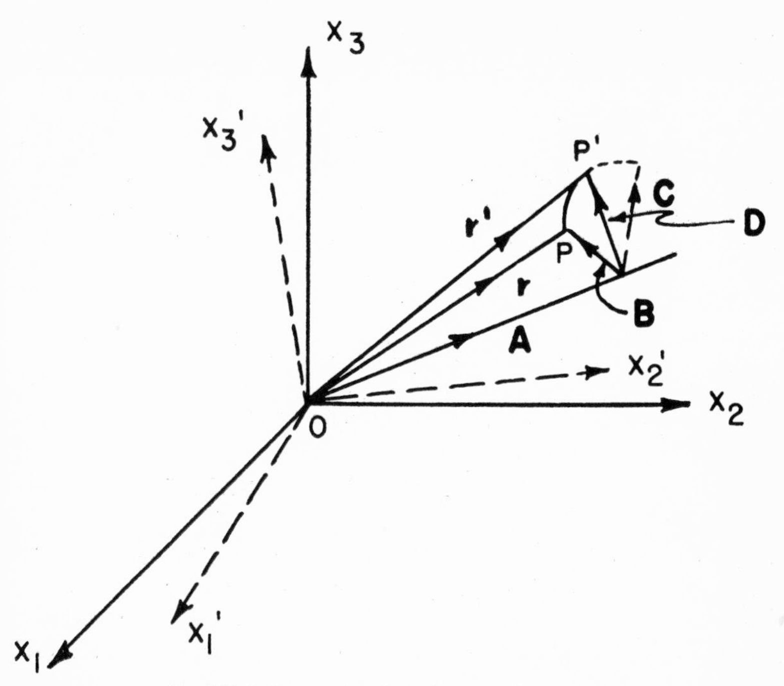

FIG. II–2. Determination of angle of rotation.

TABLE 3. Parametric representation of direction cosines.

	Axis		
Axis	x_1	x_2	x_3
x_1'	$\rho^2 + \xi^2 - \eta^2 - \zeta^2$	$2(\xi\eta + \rho\zeta)$	$2(\xi\zeta - \rho\eta)$
x_2'	$2(\xi\eta - \rho\zeta)$	$\rho^2 - \xi^2 + \eta^2 - \zeta^2$	$2(\eta\zeta + \rho\xi)$
x_3'	$2(\xi\zeta - \rho\eta)$	$2(\eta\zeta - \rho\xi)$	$\rho^2 - \xi^2 - \eta^2 + \zeta^2$

so that

$$\xi^2 + \eta^2 + \zeta^2 + \rho^2 = 1.$$

This parametric representation of Euler may be put into a more convenient form due to Rodrigues. If λ_1, λ_2, λ_3 are any three real numbers, then a corresponding ξ, η, ζ may be defined by

$$\xi = \frac{\lambda_1}{\sqrt{(1 + \lambda_k\lambda_k)}}, \quad \eta = \frac{\lambda_2}{\sqrt{(1 + \lambda_k\lambda_k)}}, \quad \zeta = \frac{\lambda_3}{\sqrt{(1 + \lambda_k\lambda_k)}},$$

where for definiteness the positive square root is chosen. The value of ρ is then

$$\rho = \frac{1}{\sqrt{(1 + \lambda_k\lambda_k)}}$$

and the cosine table may now be expressed in terms of the (λ_i).

Finally, it is of interest to calculate the (A_i) and θ corresponding to a given table (α_{ij}). It follows from Euler's representation that

$$\alpha_{11} + \alpha_{22} + \alpha_{33} = 4\rho^2 - 1,$$

so that

$$\rho = \pm \tfrac{1}{2}\sqrt{(1 + \alpha_{11} + \alpha_{22} + \alpha_{33})}.$$

Also,

$$\alpha_{23} - \alpha_{32} = 4\rho\xi, \text{ and so on,}$$

so that

$$\xi = \frac{\alpha_{23} - \alpha_{32}}{4\rho}, \quad \eta = \frac{\alpha_{31} - \alpha_{13}}{4\rho}, \quad \zeta = \frac{\alpha_{12} - \alpha_{21}}{4\rho},$$

which then determine A_i and θ. (Two solutions for θ between 0 and 360° may be found, depending on whether the + or − sign is chosen for ρ; since the sum of these two angles is 360° and since the corresponding axes of rotation are oppositely directed, the physical results are identical.)

Exercises

1. Show that the cosine table corresponding to rotation of 180° about the line passing through the origin and the point (x_1, x_2, x_3) is given by

$$\alpha_{ij} = -\delta_{ij} + 2\frac{x_i x_j}{x_k x_k}$$

2. Consider three coordinate systems, (x_i), (x_i'), and (x_i''), all with a common origin. Let $x_i' = \alpha_{ij}x_j$, $x_i'' = \alpha_{ij}'x_j'$, $x_i'' = \alpha_{ij}''x_j$. Show that

$$\alpha_{is}'' = \alpha_{ij}'\alpha_{js}.$$

3. As the orthogonal triad of Sec. I–11 moves along a space curve, it both translates and rotates in space. Show that, insofar as rotation is concerned, the instantaneous axis of rotation lies in the $(\alpha_i - \gamma_i)$ plane and that the rate of rotation is

$$\frac{d\theta}{ds} = \left(\frac{1}{T^2} + \frac{1}{R^2}\right)^{\frac{1}{2}}.$$

4. Show that the order in which two infinitesimal rotations about the origin is carried out is immaterial. (This statement does not of course hold for finite rotations.)

5. Show that any displacement of a rigid body can be obtained by first translating the body in some direction and then rotating it about this line of translation.

II–4. Vectors and Tensors

Suppose that a fluid is in motion through a certain portion of space; at any point P the fluid velocity is completely determined by a knowledge of its three components. Suppose that two coordinate systems are available. Let the velocity components at P be (V_i) and (V_i') in the (x_i) and (x_i') systems respectively. Then, just as it was previously proved that $x_i' = \alpha_{ij}x_j$, it may now be shown by the reader that

$$V_i' = \alpha_{ij}V_j$$

and

$$V_i = \alpha_{ji}V_j'.$$

The same transformation law holds for any vector, whether it is a physical quantity such as force or velocity, a geometrical quantity such as radius vector from origin, or a less easily visualized quantity such as the gradient of a scalar. For example, if

$$G_i = \frac{\partial \phi}{\partial x_i},$$

then

$$G_i' = \frac{\partial \phi}{\partial x_i'} = \frac{\partial \phi}{\partial x_k}\frac{\partial x_k}{\partial x_i'} = \alpha_{ik}G_k.$$

One of the reasons for our interest in the way in which vector components transform as the coordinate system is rotated arises from the fact that many elastic bodies do not possess preferred directions; consequently, the laws governing their behavior must be independent of the particular choice of coordinate system, and this fact leads to an a priori simplification of these laws — this simplification being based on the results of this chapter.

The foregoing transformation rule, in which each new vector component is a linear combination of the old components, is very convenient, and will be of considerable use. In fact, we shall adopt it as the definition of a vector, thus replacing the previous definition of a vector as a quantity possessing direction and magnitude. The basic reason for adopting this new definition of a vector is that it can be easily generalized to apply to more complicated physical quantities (tensors) whereas the "magnitude and direction" definition can not.

Thus we will define a vector to be a set of three quantities (called its components) possessing the property that if their values at a fixed point P in any coordinate system (x_i) are (A_i), then their values at P in any other coordinate system (x_i') are given by

$$A_i' = \alpha_{ij}A_j.$$

An equivalent statement (multiply through by α_{is}) is

$$A_i = \alpha_{ji}A_j'.$$

It must be again emphasized that all coordinate systems considered are right-handed. The reason for this may be seen by considering the cross product of two vectors; let $\mathbf{C} = \mathbf{A} \times \mathbf{B}$. Then in a right-handed coordinate system (x_i) we have

$$C_1 = A_2B_3 - A_3B_2, \text{ and so on.}$$

If we were to consider an alternative coordinate system (x_i'), we would hope that

$$C_1' = A_2'B_3' - A_3'B_2', \text{ and so on,}$$

and indeed (as will be shown) this will be the case if (x_i') system is also right-handed. Consider, however, the particular left-handed coordinate system

$$x_1' = -x_1, \; x_2' = x_2, \; x_3' = x_3.$$

Geometric visualization gives

$$C_1' = -C_1, \; A_2' = A_2, \; A_3' = A_3, \; B_2' = B_2, \; B_3' = B_3.$$

Substituting these relations into the equation for C_1 gives

$$C_1' = -(A_2'B_3' - A_3'B_2'),$$

so that an irritating negative sign has now appeared. It is in order to avoid this type of inconvenience that we restrict ourselves to right-handed systems.

So far, only scalars and vectors have been considered. However, early in the study of elasticity and electrostatics, quantities of a more complex nature were encountered. These quantities were originally called dyads. A simple example of one may be constructed by combining two vectors (A_i) and (B_i) so as to form a set of nine quantities (C_{ij}) defined by

$$C_{ij} = A_iB_j$$

(for example, $C_{23} = A_2B_3$). If it is required that the same sort of definition is to be used in all coordinate systems, then in the (x_i') system

$$C_{ij}' = A_i'B_j' = (\alpha_{is}A_s)(\alpha_{jk}B_k) = \alpha_{is}\alpha_{jk}C_{sk},$$

and the analogy to the vector transformation rule is evident. Although not all dyads can be obtained by combining two vectors as above, all dyads do have the same transformation law. A dyad is today called a second-order tensor, the name "tensor" arising from the historical association with stress (tension).

II–5. Definitions

(*a*) *A zeroth-order tensor* (or a scalar) is defined to be a single quantity dependent on position but not on coordinate system. Thus, although temperature may vary from point to point, the temperature at a fixed point in space has the same value irrespective of the coordinate system used to specify that point, and hence temperature is a scalar.

(*b*) A *first-order tensor* (or a vector) is a set of $3^1 = 3$ quantities, such that if their values at a given point are (A_i) in a coordinate system (x_i) their values (A_i') at the same point in any other coordinate system (x_i') are given by $A_i' = \alpha_{ij}A_j$.

(*c*) A *second-order tensor* is a set of $3^2 = 9$ quantities, such that if their values at a given point are (A_{ij}) in a coordinate system (x_i) their values (A_{ij}') at the same point in any other coordinate system (x_i') are given by

$$A_{ij}' = \alpha_{is}\alpha_{jk}A_{sk}.$$

It will subsequently appear that the quantities expressing the state of stress at a point in a body form a second-order tensor.

(*d*) A *third-order tensor* is a set of $3^3 = 27$ quantities, such that if their values at a given point are (A_{ijk}) in a coordinate system (x_i) their values (A_{ijk}') in any other coordinate system (x_i') are given by

$$A_{ijk}' = \alpha_{is}\alpha_{jt}\alpha_{kp}A_{stp}.$$

Tensors may be of any order; the general rule of transformation is evident from the foregoing definitions. All such tensors are called *Cartesian tensors* because of the restriction to Cartesian coordinate systems.

Example

Suppose that the nine "components" of a second-order tensor (whose *ks* component is B_{ks}) are:

$$B_{11} = 1,\ B_{12} = -1,\ B_{32} = 2,\ \text{all other } B_{ij} = 0.$$

Then if a new coordinate system (x_i') related to the (x_i) system by the cosine table given in Table 4 is introduced, the new components (B_{ij}')

TABLE 4. Direction cosines for example.

	Axis		
Axis	x_1	x_2	x_3
x_1'	$1/\sqrt{2}$	$1/\sqrt{2}$	0
x_2'	$-1/\sqrt{2}$	$1/\sqrt{2}$	0
x_3'	0	0	1

will be given by:

$$\begin{aligned} B_{11}' &= \alpha_{1k}\alpha_{1r}B_{kr} \\ &= \alpha_{11}\alpha_{11}B_{11} + \alpha_{11}\alpha_{12}B_{12} + \alpha_{13}\alpha_{12}B_{32} + 0 \\ &= \tfrac{1}{2}(1) + \tfrac{1}{2}(-1) + 0 = 0. \end{aligned}$$

Similarly, $B_{12}' = -1$, $B_{32}' = \sqrt{2}$, and so on.

II–6. Properties of Tensors

(*a*) *Transitivity.* Consider three possible coordinate systems (x_i), (x_i'), and (x_i''). Let the elements of the cosine tables relating these three systems be defined by

$$x_i' = \alpha_{ij}x_j,\ x_i'' = \alpha_{ij}'x_j',\ x_i'' = \alpha_{ij}''x_j.$$

If the components of a tensor (B_{ij}) are known in the (x_i) system, we could calculate its components (B_{ij}'') in the (x_i'') system in two ways — either directly by

$$B_{ij}'' = \alpha_{ik}''\alpha_{js}''B_{ks}$$

or indirectly by way of the (x_i') system, by

$$\begin{aligned} B_{ij}'' &= \alpha_{ip}'\alpha_{jt}'B_{pt}' \\ &= (\alpha_{ip}'\alpha_{jt}')(\alpha_{pk}\alpha_{ts}B_{ks}). \end{aligned}$$

Then the principle of transitivity states that the two results are the same, a fact which is easily proved by use of the result of Ex. 2 of Sec. II–3. The same sort of result holds for higher-order tensors.

(*b*) *Addition of tensors.* The sum or difference of two tensors of the same order is a tensor, also of the same order. The truth of this theorem can be seen by considering an example in which two second-order tensors are added. Let (A_{ij}) and (B_{ij}) be two such tensors. Let the nine quantities (C_{ij}) be defined by

$$C_{ij} = A_{ij} + B_{ij}$$

in all coordinate systems. Thus

$$C_{ij}' = A_{ij}' + B_{ij}', \text{ and so on.}$$

But the set of quantities (C_{ij}) defined in this manner forms a second-order tensor, for

$$\begin{aligned} C_{ij}' &= A_{ij}' + B_{ij}' \\ &= (\alpha_{ik}\alpha_{js})A_{ks} + (\alpha_{ik}\alpha_{js})B_{ks} \\ &= \alpha_{ik}\alpha_{js}C_{ks}, \end{aligned}$$

which is the rule for tensor transformation.

It is important to note that it cannot be determined whether or not a set of quantities is a tensor unless the values of the quantities are known in all coordinate systems, because the definition of a tensor involves the manner in which the quantities in one system are related to those in another.

(*c*) *Tensor equations.* A tensor equation which is true in one coordinate system is true in all systems. For if two tensors satisfy $A_{ij} = B_{ij}$ in the (x_i) system, define $C_{ij} = A_{ij} - B_{ij}$ in all systems. Then, by the preceding theorem, C_{ij} is a tensor. Now C_{ij} vanishes in the (x_i) system, and hence in all systems, because C_{ij}' in any system is a linear combination of the (C_{ij}).

(*d*) *Multiplication.* Consider the two tensors (A_i) and (B_{ij}). We may define a new set of quantities (C_{ijk}) by a process called tensor multiplication:

$$C_{ijk} = A_iB_{jk}.$$

It is understood that a similar rule of definition is to be used in other coordinate systems. The reader may then show that (C_{ijk}) is a third-order tensor. In general, tensor multiplication yields a new tensor whose order is the sum of the original orders.

(*e*) *Contraction.* Consider the tensor (A_{ijk}) —a set of 27 quantities. If we replace the j by a k, giving (A_{ikk}), only three quantities remain, each being the sum of three of the original components. It is easy to show that this set of three quantities is a first-order tensor. For

$$A_{ijk}' = \alpha_{ip}\alpha_{jq}\alpha_{kr}A_{pqr},$$

and therefore

$$\begin{aligned} A_{ikk}' &= \alpha_{ip}(\alpha_{kq}\alpha_{kr})A_{pqr} \\ &= \alpha_{ip}\delta_{qr}A_{pqr} \\ &= \alpha_{ip}A_{prr}, \end{aligned}$$

which is the rule for first-order tensor transformation.

(*f*) *Symmetry and skew-symmetry.* If a tensor is symmetric (or skew-symmetric) in a pair of indices in one coordinate system, then it is symmetric (or skew-symmetric) in all coordinate systems. For example, if

$$A_{ijk} = A_{ikj} \text{ in the } (x_i) \text{ system,}$$

then

$$A_{ijk}' = A_{ikj}'$$

in the (x_i') system also — a fact which follows at once because we are dealing with a tensor equation. The tensor (A_{ijk}) considered above is said to be symmetric in j and k; if $A_{ijk} = -A_{ikj}$, it is skew-symmetric in j and k. Note that any second-order tensor can be written as the sum of a symmetric and a skew-symmetric tensor, for

$$\begin{aligned} A_{ij} &= \tfrac{1}{2}(A_{ij} + A_{ji}) + \tfrac{1}{2}(A_{ij} - A_{ji}) \\ &= C_{ij} + D_{ij}, \end{aligned}$$

where C_{ij} is symmetric and D_{ij} skew-symmetric.

Exercises

1. Choose any arbitrary nontrivial set of nine numbers as the components at P of a second-order tensor in the (x_i) system. Choose two nontrivial successive rotations, resulting in (x_i') and (x_i'') systems, and verify by numerical calculation that transitivity holds.
2. Prove that $C_{ijkem} = A_{ij}B_{kem}$ is a tensor if (A_{ij}) and (B_{kem}) are.
3. Prove that if we define $\delta_{ij}' = \delta_{ij}$, then (δ_{ij}) is a second-order tensor.
4. Given a scalar ϕ, show that $(\phi_{,i})$ is a first-order tensor, that $(\phi_{,ij})$ is a second-order tensor, and that $(\phi_{,kk})$ is a scalar.
5. Given a tensor A_{ijk}, show that $(A_{ijk,s})$ is a fourth-order tensor.
6. Prove that, if we define $e_{ijk}' = e_{ijk}$, then (e_{ijk}) is a third-order tensor. (*Hint:* Write down the equation of transformation that would have to hold if e_{ijk} were a tensor, and then interpret this equation in terms of the determinant of the α_{ij}'s).
7. Show that, if (A_i) is a vector, the set of numbers

$$(C_{ij}) = \begin{pmatrix} 0 & A_3 & -A_2 \\ -A_3 & 0 & A_1 \\ A_2 & -A_1 & 0 \end{pmatrix}$$

forms a skew-symmetric second-order tensor.

These various results have some interesting consequences. For example, given two vectors (A_i) and (B_i), define the set (C_i) by

$$C_i = e_{ijk}A_jB_k.$$

Then (C_i) is a vector, because it is the result of contracting a tensor product.

II–7. Tests for Tensor Character

It may occasionally be inconvenient to apply the definition directly in order to determine whether or not a set of quantities is a tensor; some alternative tests for tensor character exist, and will be demonstrated by two simple examples. Tests for higher-order tensors are exactly similar.

(*a*) If (A_i) is a set of three quantities possessing the property that A_iB_i is a scalar for every vector (B_i), then (A_i) is a vector also; for

$$A_i'B_i' = A_iB_i,$$

that is,

$$(A_i'\alpha_{ij} - A_j)B_j = 0$$

and since (B_i) is arbitrary this implies that

$$A_i'\alpha_{ij} = A_j,$$

which is equivalent to the rule of tensor transformation.

(*b*) If (C_{ij}) is a set of nine quantities possessing the property that $C_{ij}A_iB_j$ is a scalar for all vectors (A_i), (B_i), then C_{ij} is a second-order tensor. The proof is similar.

Exercises

1. Given a set of nine quantities (B_{ij}), show that, if $B_{ij}C_j$ is a vector for all vectors (C_j), then (B_{ij}) is a second-order tensor.
2. Given a set of nine quantities (B_{ij}), show that, if (B_{ij}) is symmetric in all coordinate systems and if $B_{ij}C_iC_j$ is a scalar for all vectors (C_i), then (B_{ij}) is a tensor. Why is the symmetry condition required?

II–8. Isotropic Tensors

An isotropic tensor is one whose components are the same in all coordinate systems. Examples are δ_{ij} and e_{ijk}, as indicated in Exercises 3 and 6 of Sec. II–6.

Clearly every scalar is an isotropic tensor. There is, however, no nontrivial isotropic vector. For such a vector would have to satisfy

$$A_i = A_i' = \alpha_{ij}A_j$$

for all possible coordinate transformations. In particular, for a 180° rotation about the x_1-axis ($x_1' = x_1$, $x_2' = -x_2$, $x_3' = -x_3$), the equation becomes

$$A_1 = A_1,\ A_2 = -A_2,\ A_3 = -A_3.$$

Hence $A_2 = A_3 = 0$; similarly $A_1 = 0$.

By similar choices of special coordinate transformations, it may be shown that any second-order isotropic tensor must be of the form of a constant times (δ_{ij}), and any third-order isotropic tensor must be of the form of a constant times (e_{ijk}). The important case of a fourth-order isotropic tensor is a little more difficult; we will consider it in some detail.

Theorem: The most general fourth-order isotropic tensor has the form

$$A_{ijkm} = \alpha\delta_{ij}\delta_{km} + \beta\delta_{ik}\delta_{jm} + \gamma\delta_{im}\delta_{jk},$$

where α, β, γ are constants.

Proof: The hypothesis requires that

$$A'_{ijkm} = \alpha_{ir}\alpha_{js}\alpha_{kt}\alpha_{mn}A_{rstn}.$$

Since each of the subscripts (i, j, k, m) may have three values, there are $3^4 = 81$ components altogether. Consider first any component which contains the index 1 only once. For the coordinate transformation associated with a 180° rotation about the x_1-axis, only one term appears on the right-hand side of the equation, namely, the negative of the left-hand side. Hence any component which contains the index 1 only once — and similarly any component which contains any index only once — must vanish. Now for the coordinate transformation ($x_1' = x_2$, $x_2' = x_3$, $x_3' = x_1$), the equation yields

$$A_{1122} = A_{2233},\ A_{1212} = A_{2323},\ A_{1111} = A_{2222}, \text{ and so on.}$$

By making use of similar special transformations the reader may show that

$$\begin{aligned}
A_{1122} &= A_{2211} = A_{1133} = A_{3311} = A_{2233} = A_{3322} = \alpha, \text{ say;}\\
A_{1212} &= A_{2121} = A_{1313} = A_{3131} = A_{2323} = A_{3232} = \beta, \text{ say;}\\
A_{1221} &= A_{2112} = A_{1331} = A_{3113} = A_{2332} = A_{3223} = \gamma, \text{ say;}\\
A_{1111} &= A_{2222} = A_{3333} = \alpha + \beta + \gamma;
\end{aligned}$$

and as previously proved all other A_{ijkl} vanish. But a little thought will now show that this means that A_{ijkl} must indeed have the required form.

(In general, it may be shown that any even-order isotropic tensor has a form precisely analogous to the fourth-order case above — that is, all possible combinations of δ_{ij} appear. The most general odd-order isotropic tensor has a form analogous to that for the fifth order:

$$\alpha e_{ijk}\delta_{mn} + \beta e_{ijm}\delta_{kn} + \gamma e_{ijn}\delta_{km} + \delta e_{ikm}\delta_{jn} + \cdots)$$

Exercises

1. Prove that there is no pair of vectors, (A_i) and (B_i), such that

$$\delta_{ij} = A_i B_j.$$

2. Show that an arbitrary second-order tensor (C_{ij}) may be written in the form

$$C_{ij} = \alpha\delta_{ij} + B_{ij}$$

where $B_{ii} = 0$.

3. Prove that

(a) $e_{ijk}\delta_{mn} = e_{mjk}\delta_{in} + e_{imk}\delta_{jn} + e_{ijm}\delta_{kn}$,

(b) $$e_{ijk}e_{mnr} = \begin{vmatrix} \delta_{im} & \delta_{in} & \delta_{ir} \\ \delta_{jm} & \delta_{jn} & \delta_{jr} \\ \delta_{km} & \delta_{kn} & \delta_{kr} \end{vmatrix}.$$

III

Stress

III–1. Definition of Stress Tensor

The discussion in this chapter will apply to any material body that can be treated as a continuum.

Consider a plane area element drawn through a point P inside the body (Fig. III–1) and let n_i be a unit normal vector originating at P. (There

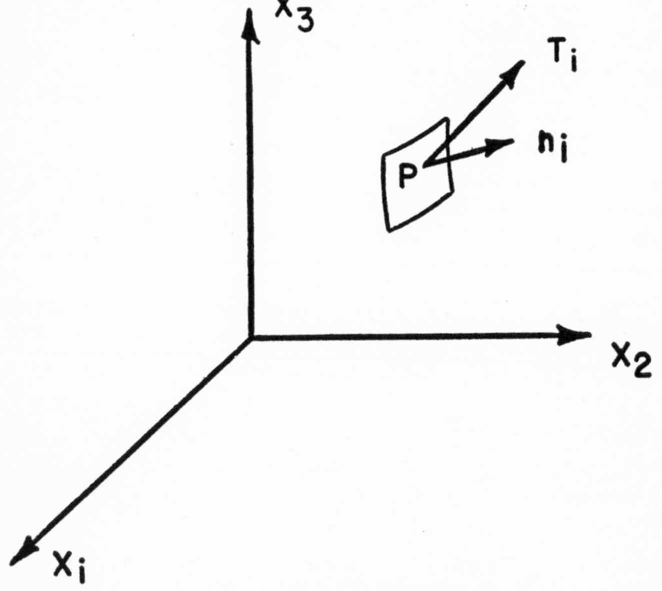

FIG. III–1. Stress on area element.

are of course two such vectors, pointing in opposite directions. Either may be chosen, but the choice thereafter must not be altered.) Denote by

M_+ the material immediately adjacent to that side of the area element toward which n_i points and by M_- the material immediately adjacent to the other side of the area element. In general, M_+ will exert a certain vector force on M_-. Now let the area element shrink uniformly in size (insisting however that the linear dimensions of the area element remain large in comparison with molecular spacings, so as to stay within the domain of continuum mechanics). Then it is physically plausible to assume that the ratio of the above-mentioned vector force to the size of the area element approaches a definite limit; this limiting ratio is called the *stress vector* T_i. The dimensions of T_i are force per unit area.

In addition to a vector force, M_+ may well exert a vector moment on M_-; it is conventional to assume that the ratio of this moment to the area of the element approaches zero as the limiting process described above is carried out. At first sight this may appear strange — at least for magnetized or electrically polarized bodies — for it is easy to visualize M_- as consisting, say, of elemental magnets lined up in some common direction; if then the M_+ elemental magnets generate a magnetic field tending to rotate the M_- magnets, there certainly will be a resultant nonzero moment per unit area exerted by M_+ on M_-. However, unless there is an abrupt discontinuity in the elemental magnet orientation across the area element, such effects should be negligible. (Remember that only a few layers of molecules are involved in either M_+ or M_-. Although it is true that the remainder of the body could produce a substantial magnetic field tending to rotate the M_- magnets, such torques are considered body moments rather than surface traction moments.) Since such discontinuities are not permissible in a physical continuum, it is indeed reasonable to assume the limiting value of moment intensity to be zero. Of couse, the ultimate criterion for the validity of the two assumptions (that a definite limit T_i exists, and that the limiting moment intensity is zero) is obtained by experimentally testing deductions based on these assumptions; on this basis, these two assumptions must be regarded as very well verified indeed. Philosophically, one may remark that, as long as a simple physically plausible hypothesis continues to yield correct results, there is no point in searching for complications.

Consider next the problem of completely specifying the state of stress at the point P. One way of doing this would be to list the values of T_i corresponding to various area elements, that is, to various n_i. It turns out, however, that these stress vectors are related to one another; in fact, the value of T_i for any n_i may be calculated once the stresses are known for those three area elements whose normals are in the directions of the coordinate axes.

Denote by τ_{ij} the jth component of the stress vector acting on an area element (at P) whose normal is in the direction of the positive x_i-axis (Fig. III–2). For example, a metal rod oriented to lie along the x_2-axis

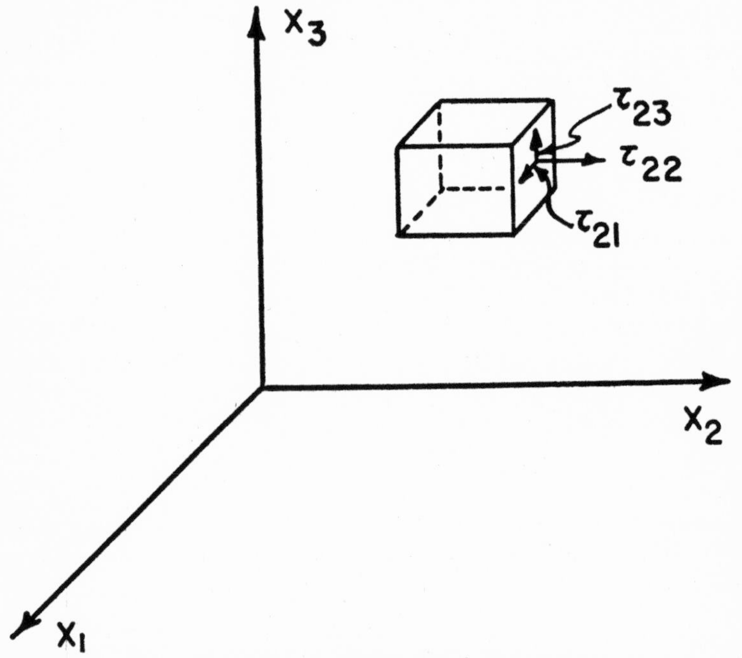

FIG. III–2. Typical components of stress tensor.

and subjected to a compressive end load of 16 lb/in.² would have $\tau_{22} = -16$ lb/in.², all other $\tau_{ij} = 0$ (and subsequent elasticity analysis will prove that this is the only possible stress state in the interior of the bar). The reason for the negative sign is that M_+ here exerts a compressive force on M_-, so that the stress vector points in the direction of the negative x_2-axis.

Consider now an arbitrary area element with unit normal vector n_i (Fig. III–3); what is T_i for this element in terms of the τ_{ij}? Construct a tetrahedon with apex at P as shown (this tetrahedon will be allowed to approach zero size; in the limit, the area element will pass through P). Let the material density be ρ, the body force (such as gravity) per unit

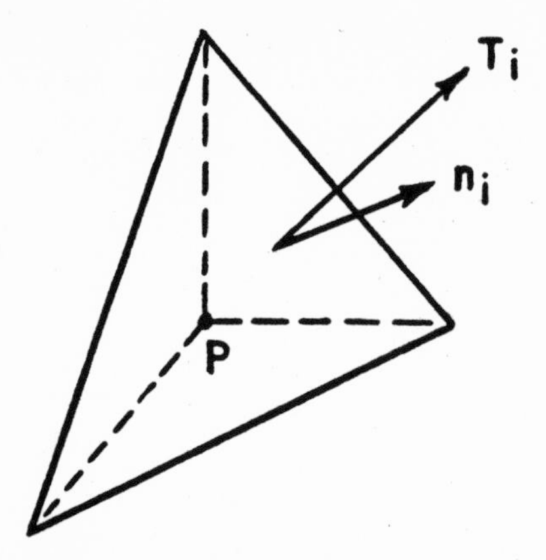

FIG. III-3. Relation between stress vector and stress tensor components.

mass be F_i, and let the volume element shown be undergoing an acceleration a_i. If the size of the area element is ΔS and if the perpendicular distance from P to the area element is h (so that the volume of the tetrahedon is $\frac{1}{3}h\,\Delta S$), then Newton's law for the acceleration in the x_i-direction gives

$$F_i(\rho \cdot \tfrac{1}{3}h \cdot \Delta S) + T_i(\Delta S) = \tau_{1i}(n_1\,\Delta S) + \tau_{2i}(n_2\,\Delta S) + \tau_{3i}(n_3\,\Delta S) + a_i(\tfrac{1}{3}h\rho\,\Delta S)$$

If h is allowed to approach zero, the equation becomes

$$T_i = \tau_{ji}n_j, \qquad \text{(III-1)}$$

and it therefore does follow that T_i for any n_i may be calculated from a knowledge of the nine basic quantities (τ_{pq}).

These nine quantities form a second-order tensor — called the *stress tensor* at P — as may be verified at once by noting that T_i is a vector for arbitrary n_i, so that the test for tensor character of (Sec. II-7) is applicable. Some useful conclusions may be drawn from this fact:

1. If the stress tensor at P is τ_{ij} in the x_i system, then it is τ_{ij}' in the x_i'-system, where

$$\begin{aligned} \tau_{ij}' &= \alpha_{ip}\alpha_{jq}\tau_{pq}, \\ \tau_{ij} &= \alpha_{pi}\alpha_{qj}\tau_{pq}'. \end{aligned} \qquad \text{(III-2)}$$

2. τ_{jj} is a scalar, and has the same value in all coordinate systems.

The stress vector T_i acting on the area element n_i at P may be resolved into normal and shear components. The magnitude of the normal stress is clearly $T_j n_j$, so that the normal stress vector N_i is

$$N_i = T_j n_j n_i = \tau_{sj} n_s n_j n_i.$$

The shear stress vector is

$$S_i = T_i - N_i.$$

III–2. Equations of Equilibrium

1. Forces

With the foregoing notation, Newton's law applied to an arbitrary portion of the material (this portion having volume V and surface S) gives for the x_i-component of acceleration

$$\int \rho F_i \, dV + \int T_i \, dS = \int \rho a_i \, dV.$$

The second integral becomes

$$\int \tau_{ji} n_j \, dS = \int \tau_{ji,j} \, dV$$

by Gauss's theorem (Sec. I–13); consequently

$$\int (\rho F_i + \tau_{ji,j} - \rho a_i) \, dV = 0$$

for any such volume V. But this implies that the integrand must be zero, for if there were some point at which the integrand were not zero, but, say, positive, then continuity would require the integrand to be positive over some small region surrounding this point; but the integral could then not vanish if V were chosen to consist of this small region alone. It follows that

$$\tau_{ji,j} + \rho F_i = \rho a_i \tag{III–3}$$

everywhere.

2. Moments

The ith component of the moment about the origin of any force G_i applied at a point with coordinates x_i is given by $e_{ijk} x_j G_k$. For the volume V, the resultant moment of all applied forces must equal the net moment of the mass acceleration, so that

$$\int e_{ijk} x_j F_k \rho \, dV + \int e_{ijk} x_j T_k \, dS + \int M_i \rho \, dV = \int e_{ijk} x_j a_k \rho \, dV,$$

where M_i is the body moment per unit mass (resulting perhaps from an external magnetic field acting on a magnetized medium). Replacing T_k by $\tau_{rk}n_r$, using the divergence theorem as in (*1*) above, and incorporating Eq. (III–3) gives eventually

$$e_{ijk}\tau_{jk} + \rho M_i = 0.$$

Body moments rarely occur in elasticity, so that it is sufficiently general to set $M_i = 0$ in the preceding equation, which then reduces to

$$\tau_{ij} = \tau_{ji}, \qquad \text{(III–4)}$$

so that the stress tensor is symmetric. This means that Eqs. (III–1) and (III–3) may be rewritten somewhat more conventionally:

$$T_i = \tau_{ij}n_j, \qquad \text{(III–1)}$$

$$\tau_{ij,j} + \rho F_i = \rho a_i. \qquad \text{(III–3)}$$

III–3. Principal Axes

At a point P inside the material, the stress on an area element n_i is given by Eq. (III–1). Is there any area element at P such that T_i is perpendicular to the area element, that is, such that $T_i = \lambda n_i$? For such an element, we would have

$$(\tau_{ij} - \lambda\delta_{ij})n_j = 0, \qquad \text{(III–5)}$$

which is a set of three linear homogeneous equations for (n_1, n_2, n_3). This set possesses solutions if and only if the determinant of the coefficients vanishes:

$$D = \begin{vmatrix} \tau_{11} - \lambda & \tau_{12} & \tau_{13} \\ \tau_{21} & \tau_{22} - \lambda & \tau_{23} \\ \tau_{31} & \tau_{32} & \tau_{33} - \lambda \end{vmatrix} = 0, \qquad \text{(III–6)}$$

so that this requirement determines the value of λ. There are in general three roots, $\lambda^{(1)}$, $\lambda^{(2)}$, and $\lambda^{(3)}$; since the basic equation was $T_i = \lambda n_i$, these three possible values of λ are the three possible magnitudes of normal stress corresponding to zero shear stress.

Assume that $\lambda^{(1)}, \lambda^{(2)}, \lambda^{(3)}$ are all different (the special case in which two or more coincide may subsequently be treated as a limiting case). If then λ in Eqs. (III–5) is set equal to, say, $\lambda^{(1)}$, Eq. (III–6) implies from linear-equation theory that at most two of the three equations (III–5) can be independent. It will in fact subsequently appear (as a result of the assumption that the three λ-roots are all different) that *exactly* two of the three equations are independent; however, in the meantime we need only the resulting fact that, whether two or only one of Eqs. (III–5) are independent, at least one solution $n_i^{(1)}$ satisfying Eqs. (III–5) and also

$$n_i n_i = 1$$

exists. Similarly, an $n_i^{(2)}$ corresponding to $\lambda^{(2)}$ and an $n_i^{(3)}$ corresponding to $\lambda^{(3)}$ may be found.

An example may be worth while. Suppose that the stress state at P is given by $\tau_{11} = -10$ lb/in.^{2}, $\tau_{12} = 9$ lb/in.^{2}, $\tau_{13} = 5$ lb/in.^{2}, $\tau_{22} = \tau_{23} = 0$, $\tau_{33} = 8$ lb/in.^{2}. Then Eq. (III–6) becomes

$$(-10 - \lambda)(-\lambda)(8 - \lambda) - (5)(5)(-\lambda) - (9)(9)(8 - \lambda) = 0,$$

which has the three roots

$$\begin{aligned} \lambda^{(1)} &= 4 \text{ lb/in.}^2, \\ \lambda^{(2)} &= 10.08 \text{ lb/in.}^2, \\ \lambda^{(3)} &= -16.08 \text{ lb/in.}^2. \end{aligned} \tag{III–7}$$

As a check, the sum of these values equals $\tau_{ii} = -2$. To find the three components $n_i^{(1)}$ corresponding to $\lambda^{(1)}$ it is necessary to solve the set

$$\begin{aligned} (-10 - 4)n_1^{(1)} + (9)n_2^{(1)} + (5)n_3^{(1)} &= 0, \\ (9)n_1^{(1)} + (0 - 4)n_2^{(1)} + (0)n_3^{(1)} &= 0, \\ (5)n_1^{(1)} + (0)n_2^{(1)} + (8 - 4)n_3^{(1)} &= 0. \end{aligned}$$

The first of these equations may be obtained by adding $(-\frac{9}{4})$ times the second equation to $(\frac{5}{4})$ times the third equation, and so the first equation is redundant and may be discarded. The remaining two equations have the solution

$$\begin{aligned} n_2^{(1)} &= \tfrac{9}{4} n_1^{(1)}, \\ n_3^{(1)} &= -\tfrac{5}{4} n_1^{(1)}. \end{aligned}$$

Using now the condition that $n_i^{(1)} n_i^{(1)} = 1$ gives the final solution as

$$n_i^{(1)} = \left(\frac{4}{\sqrt{122}}, \frac{9}{\sqrt{122}}, \frac{-5}{\sqrt{122}}\right)$$

or as its negative,

$$n_i^{(1)} = \left(\frac{-4}{\sqrt{122}}, \frac{-9}{\sqrt{122}}, \frac{5}{\sqrt{122}}\right).$$

Thus an area element perpendicular to this direction experiences only a normal stress — whose value is in fact 4 lb/in.^{2} tensile. The other directions, $n_i^{(2)}$ and $n_i^{(3)}$, could be found similarly.

Return now to the general problem. It will first be proved that the unit normals corresponding to different λ-roots are orthogonal. For if $n_i^{(1)}$ corresponds to $\lambda^{(1)}$ and $n_i^{(2)}$ to $\lambda^{(2)}$, then multiplying

$$\tau_{ij}n_j^{(1)} = \lambda^{(1)}n_i^{(1)}$$

by $n_i^{(2)}$ gives

$$\tau_{ij}n_i^{(2)}n_j^{(1)} = \lambda^{(1)}n_i^{(1)}n_i^{(2)}.$$

Similarly,

$$\tau_{ij}n_i^{(1)}n_j^{(2)} = \lambda^{(2)}n_i^{(2)}n_i^{(1)}.$$

Because τ_{ij} is symmetric, the left-hand sides of these two equations are equal, and therefore

$$(\lambda^{(1)} - \lambda^{(2)})n_i^{(1)}n_i^{(2)} = 0.$$

Since $\lambda^{(1)}$ and $\lambda^{(2)}$ are different by hypothesis, we conclude that $n_i^{(1)}$ and $n_i^{(2)}$ are orthogonal, thus proving the theorem.

It was previously remarked that when λ was set equal to a root of Eq. (III–6) exactly two of Eqs. (III–5) would be independent. This fact follows now from the orthogonality of the n_i vectors; if for $\lambda^{(1)}$, say, only one of Eqs. (III–5) were independent, there would then be considerable choice for $n_i^{(1)}$, and not all such vectors could be orthogonal to a definite pair $n_i^{(2)}$, $n_i^{(3)}$.

It has up till now been implicitly assumed that all λ-roots and corresponding n_i vectors are real; this fact will now be proved by showing that the supposition that $\lambda^{(1)}$, say, is complex leads to a contradiction. For if $\lambda^{(1)}$ is complex, then taking complex conjugates of

$$\tau_{ij}n_j^{(1)} = \lambda^{(1)}n_i^{(1)}$$

gives

$$\tau_{ij}(n_j^{(1)})^* = (\lambda^{(1)})^*(n_i^{(1)})^*,$$

which may be multiplied by $n_i^{(1)}$ to give

$$\tau_{ij}n_i^{(1)}(n_j^{(1)})^* = (\lambda^{(1)})^*n_i^{(1)}(n_i^{(1)})^*.$$

But multiplying the original equation by $(n_i^{(1)})^*$ gives

$$\tau_{ij}n_j^{(1)}(n_i^{(1)})^* = \lambda^{(1)}n_i^{(1)}(n_i^{(1)})^*$$

and comparison with the last equation then yields

$$[\lambda^{(1)} - (\lambda^{(1)})^*]n_i^{(1)}(n_i^{(1)})^* = 0.$$

Since $n_i^{(1)}(n_i^{(1)})^*$ is a sum of squares of real numbers, it cannot be zero; hence $\lambda^{(1)}$ must equal its conjugate and so cannot be complex.

To summarize, it has been shown that the three λ-roots of Eq. (III–6) are real; further, if these λ-roots are different, then there exist three mutually orthogonal directions at P such that area elements perpendicular to these directions experience only normal stress — these normal stresses being in fact the three λ-roots. The three directions are called *principal directions* at P, and the corresponding normal stresses are called *principal stresses.* The reader may find it worth while to complete the numerical example above by calculating $n_i^{(2)}$ and $n_i^{(3)}$ so as to verify that $n_i^{(1)}$, $n_i^{(2)}$, and $n_i^{(3)}$ are perpendicular to one another.

A right-handed coordinate system may be so oriented as to line up with the principal directions at P; such a coordinate system is called a *principal-axis system* for the stress state at P. If τ_{ij}' denotes the stress state at P in the principal-axis system, then all off-diagonal terms in τ_{ij}' must vanish, for there are no shear stresses on area elements perpendicular to the principal directions. Note incidentally that for fixed principal directions there are 24 different ways of orienting a principal-axis system so that the axes coincide with the principal directions.

It remains to consider the situation where two (or three) of the λ-roots coincide. Suppose, for example, that $\lambda^{(1)} = \lambda^{(2)} \neq \lambda^{(3)}$. If the values of the τ_{ij} are in imagination altered very slightly so as to make $\lambda^{(1)}$ and $\lambda^{(2)}$ differ by a very small amount, then the previous theory applies to this modified stress state so that there are then three mutually orthogonal principal directions. If the modified τ_{ij} are allowed to approach the actual τ_{ij}, it follows that in the limit it must still be possible to find three mutually orthogonal principal directions; consequently the possibility of equality between principal stresses does not introduce any complications in this respect at least. It may, however, happen — because of the arbitrariness of the foregoing modifications of the τ_{ij} — that the set of three mutually perpendicular principal directions need not be unique. That such lack of uniqueness does in fact occur will be shown in the next section; for the present, however, our main interest is in the fact that, *regardless of the stress state, it is always possible to find at least one principal axis system. The corresponding principal stresses are the three real roots of Eq. (III–6).*

III–4. Some General Properties of Stress

1. Let τ_{ij} be the stress state at P with respect to some arbitrary coordinate system x_i. Let x_i' be a principal-axis system at P; denote the stress state in the x_i' system by τ_{ij}'. Consider now some particular area

element whose unit normal vector is n_i in the original system and n_i' in the principal-axis system. Then the two equations

$$T_i = \tau_{ij} n_j,$$
$$T_i' = \tau_{ij}' n_j'$$

are equivalent; the second, however, is often more useful for it reduces to

$$T_1' = \tau_{11}' n_1',$$
$$T_2' = \tau_{22}' n_2',$$
$$T_3' = \tau_{33}' n_3',$$

since there are no shear components of τ_{ij}'. Suppose, for example, that two of the principal stresses — say τ_{11}' and τ_{22}' — are equal. Then

$$\frac{T_1'}{T_2'} = \frac{n_1'}{n_2'},$$

so that the stress vector on any area element for which $n_3' = 0$ is normal to that area element; moreover, the magnitude of the stress vector on such an area element is τ_{11}'. This means that any coordinate system, one of whose axes coincides with x_3', is a principal-axis system. Similarly, if all principal stresses are equal, then *any* coordinate system is a principal-axis system.

2. A second example of the use of the principal-axis system for calculations is afforded by the problem of determining the maximal normal stress on an area element through P. The normal stress magnitude for an area element n_i' is

$$N = \tau_{ij}' n_i' n_j'$$
$$= \tau_{11}'(n_1')^2 + \tau_{22}'(n_2')^2 + \tau_{33}'(n_3')^2,$$

and, since

$$(n_1')^2 + (n_2')^2 + (n_3')^2 = 1,$$

it follows that the greatest and least normal stresses are respectively the greatest and least of the principal stresses. Thus, in the example of Sec. III–3, no area element can be found for which the tensile normal stress exceeds 10.08 lb/in.^2 nor one for which the compressive normal stress exceeds 16.08 lb/in.^2. A consequence of this theorem is the statement that, if a stress state is obtained by superposing two other stress states, then the maximal principal stress is not greater than the sum of the individual maximal principal stresses.

3. There are a number of combinations of the τ_{ij} which are not altered by a rotation of the coordinate system. One easy way of finding such

invariants is to use the fact that τ_{ij} is a tensor, so that any scalar quantity (no free indices) constructed out of τ_{ij} must be an invariant. Some such invariants (together with their values in terms of the principal stresses τ_1, τ_2, τ_3) are:

$$P_1 = \tau_{ii} = \tau_1 + \tau_2 + \tau_3,$$
$$P_2 = \tau_{ij}\tau_{ji} = \tau_1^2 + \tau_2^2 + \tau_3^2,$$
$$P_3 = \tau_{ij}\tau_{jk}\tau_{ki} = \tau_1^3 + \tau_2^3 + \tau_3^3.$$

Higher-order invariants may be constructed similarly. Another method is to use the e_{ijk} symbol:

$$Q_1 = e_{ijk}e_{ijr}\tau_{kr} = 2P_1,$$
$$Q_2 = e_{ijk}e_{ist}\tau_{js}\tau_{kt} = P_1^2 - P_2,$$
$$Q_3 = e_{ijk}e_{rst}\tau_{ir}\tau_{js}\tau_{kt} = 6\tau_1\tau_2\tau_3 = P_1^3 - 3P_1P_2 + 2P_3,$$

and so forth. The reason for the interest in these Q invariants is that they arise in the expansion of the determinant in Eq. (III–6), which may be written

$$(\lambda - \tau_1)(\lambda - \tau_2)(\lambda - \tau_3) = 0,$$

that is,

$$\lambda^3 - \lambda^2(\tfrac{1}{2}Q_1) + \lambda(\tfrac{1}{2}Q_2) - (\tfrac{1}{6}Q_3) = 0.$$

Notice that $\frac{1}{6}Q_3$ is the value of the determinant of the τ_{ij} and that $\frac{1}{2}Q_2$ is the sum of the principal two-rowed minors of this determinant.

4. An important property of the stress state at a point P may be obtained by considering two area elements passing through P; let these elements have unit normals $n_i^{(1)}$, $n_i^{(2)}$ and experience stress vectors $T_i^{(1)}$, $T_i^{(2)}$ respectively. Then the "projection" of the first stress vector on the second unit normal vector is equal to the projection of the second stress vector on the first unit normal vector; that is,

$$T_i^{(1)}n_i^{(2)} = T_i^{(2)}n_i^{(1)}.$$

The proof of this projection theorem follows from the substitution of T_i in terms of τ_{ij} and use of the symmetry of τ_{ij}.

Suppose now that the state of stress were such that one particular area element A_1 passing through P was stress free, that is, that $T_i = 0$ for the element A_1. Then use of the projection theorem shows at once that the stress vector on any other area element through P must be parallel to the plane of A_1. Conversely, if it is known that the stress vector on any area element is parallel to the plane of A_1, it will follow that A_1 is stress free (actually, it is only necessary to know that the stress vectors on three noncollinear area elements are in the same plane). Such a stress state is said to be *plane*. It is easily shown that a necessary and sufficient condition that a state of stress be plane is that one of the principal stresses be zero (that is, that the determinant of the τ_{ij} vanishes).

If there are two different stress-free elements through P, then the stress vector on any other area element must be parallel to each of these and so to their line of intersection; then the stress state is *linear*. Further, the stress vector on any area element containing this line of intersection also vanishes. Again, the converse result holds; also, a necessary and sufficient condition that a state of stress be linear is that two of the principal stresses be zero. (These remarks about principal stresses imply that any stress state may be written as the sum of a plane state of stress and a linear state of stress.)

Finally, it is easily shown that as an area element at P is rotated about an axis through P the normal stress varies at a rate proportional to the shear component perpendicular to that axis.

5. In an x_i coordinate system, let the stress state at P be τ_{ij}. If some coordinate system x_i'' can be found such that $\tau_{11}'' = \tau_{22}'' = \tau_{33}'' = 0$, the state of stress at P is termed *pure shear*. Since $\tau_{jj}'' = 0$, it follows that a necessary condition for pure shear is that $\tau_{ii} = 0$.

This condition is also sufficient. For if $\tau_{11} + \tau_{22} + \tau_{33} = 0$, then at least one of these stresses — say τ_{11} — is positive, and at least one — say τ_{22} — is negative. As an area element initially perpendicular to the x_1-axis is rotated about a line parallel to the x_3-axis until it is perpendicular to the x_2-axis, the normal stress on it changes from positive to negative; consequently there must by continuity be an intermediate orientation of the element for which the normal stress is zero. Denote the unit normal vector to this intermediate area element by n_i. Now choose two more directions such that the combination of these two and n_i form an orthogonal triad emanating from P. This triad forms a permissible coordinate system, in which the sum of the main-diagonal terms of the stress tensor is still zero; that is, the sum of the normal stresses on area elements perpendicular to the two new directions is zero and so again one stress must be tensile and the other compressive. Rotate these two area elements as a unit about the n_i-axis; clearly there will be one orientation where both normal stresses vanish (if one does, so must the other), so that the result is proved. Since the coordinate system one starts with is arbitrary, it is in fact clear that there are an infinite number of suitable x_i''-coordinate systems.

Continuity arguments of the foregoing kind are often useful in elasticity; it is, for example, possible to deduce the existence of a principal-axis system by very similar means.[1]

Since any stress state τ_{ij} may be written as

$$\tau_{ij} = \tau_{ij}' + \tfrac{1}{3}\delta_{ij}\tau_{kk}$$

(where τ_{ij}' is defined by this equation), it is seen that any stress state may be decomposed into two stress states, one of which represents pure shear and the other of which represents hydrostatic tension.

6. By changing to a principal-axis system for one of the stress states, it is easily shown that the two stress states τ_{ij} and t_{ij} have a principal-axis system in common if and only if

$$\tau_{ij}t_{jk} = t_{ij}\tau_{jk}.$$

Because of symmetries, there are only three independent equations. In the special case of two dimensions, there is only one independent equation:

$$(\tau_{11} - \tau_{22})t_{12} = (t_{11} - t_{22})\tau_{12}.$$

It is not necessary that t_{ij} be another stress tensor; it could be any symmetric tensor — such as the strain tensor of Chapter IV — and the condition for coincidence of principal axes would remain unaltered.

III–5. Geometric Properties of Stress

1. The simplest geometric representation of the stress state at a point P is obtained by treating the three principal stresses τ_1, τ_2, τ_3 as coordinates of a point in three-dimensional stress-space (Fig. III–4). Any two stress states at P which differ in the location of their principal axes but not in their principal stress values would then be represented by the same point, a fact which implies that with this kind of stress space one is interested primarily in the geometry of stress and not in the orientation of the stress state with respect to the material of the body. It is interesting to note that any two stress states which differ only by hydrostatic tension lie on a line which makes equal angles with the three axes; as the repre-

[1] Any area element at P has a shear stress vector which may be projected onto a sphere surrounding P. If this is done for all area elements, a continuous vector field is obtained on the sphere. Such a field must always have a vector of zero length; see D. Hilbert and S. Cohn-Vossen, *Geometry and the imagination* (Chelsea, New York, 1952), p. 325. Consequently at least one area element A at P has zero shear stress. Choose two area elements perpendicular to each other and to A; their shear stresses must be parallel to the plane of A, equal to each other in magnitude, and point either toward or away from each other. As this pair of area elements is rotated (about an axis perpendicular to A), each shear must therefore vanish for some orientation.

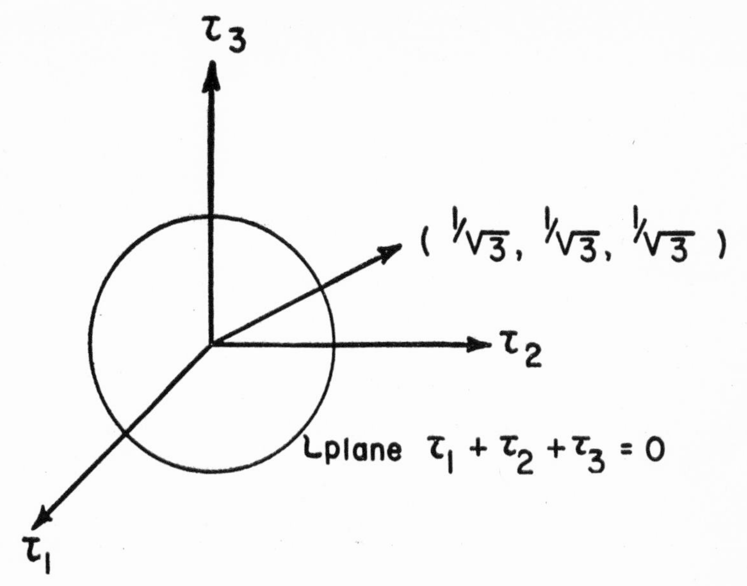

FIG. III–4. Stress space.

sentative point moves along this line until it encounters the plane $\tau_1 + \tau_2 + \tau_3 = 0$, the resulting point represents a pure shear state which differs from the original state only by a hydrostatic tension. The corresponding stress state may be written

$$S_{ij} = \tau_{ij} - \tfrac{1}{3}\delta_{ij}\tau_{kk},$$

so that $S_{ii} = 0$. The quantity S_{ij} defined by this equation is termed the *stress deviator*. Note that S_{ij} and τ_{ij} have the same principal axes.

It is also possible to consider the six independent τ_{ij} as positional coordinates in a six-dimensional space. This obviates the necessity of also specifying principal axes, but suffers from the fact that shear and normal stresses are treated on the same basis, which is rather unsymmetrical. It has been pointed out by Busemann that there exist six directions from a point which are symmetrically oriented in space, namely, the perpendiculars from the point onto the faces of a regular twelve-sided solid whose center coincides with the point, the angle between adjacent directions being about 77°, so that a very symmetrical way of describing the stress state at a point is to specify the normal stresses on area elements perpendicular to these directions. These six normal stresses would then be particularly suitable for use as coordinates in six-dimensional space.

2. Let the given stress state at P be τ_{ij}. Construct a three-dimensional space with orthogonal Cartesian axes (y_1, y_2, y_3), and consider the quadric surface

$$\tau_{ij}y_iy_j = C, \tag{III–8}$$

where the sign of the constant C is so chosen as to make the surface real. In general, the surface will consist of a number of distinct portions (sheets), C being positive for some sheets and negative for others. If an alternative (rotated) coordinate system $y_i' = \alpha_{ij}y_j$ were used (see Chapter II), then the equation of the surface would become

$$\tau_{ij}\alpha_{si}\alpha_{tj}y_s'y_t' = C.$$

If we think of the original y_i system as parallel to the physical x_i-axes, and the y_i'-system as parallel to the physical principal axes x_i', then

$$\alpha_{si}\alpha_{tj}\tau_{ij} = \tau_{st}',$$

where τ_{st}' are the principal stresses; consequently, the previous equation becomes

$$\tau_{st}'y_s'y_t' = C,$$

and now the left-hand side involves squares only, since $\tau_{st}' = 0$ for $s \neq t$. Thus there exists a y_i'-coordinate system, called the principal-axis system of the quadric surface, such that the equation of the surface takes a very simple form; this fact allows us to easily visualize the various possible types of quadric surface represented by Eq. (III–8).

Return now to Eq. (III–8), and require that the y_i-axes be parallel to the physical x_i-axes. Let A be any area element at P, and let its unit normal vector be n_i. Then the normal stress on A is

$$N = \tau_{ij}n_in_j,$$

so that, if we draw a line parallel to n_i between the origin of the y_i-system and the quadric surface, the value of N will be inversely proportional to the square of the length of this line.

Also, the gradient of the space function $\tau_{kj}y_ky_j$ is perpendicular to any surface where $\tau_{kj}y_ky_j$ is a constant, so that the vector

$$(\tau_{kj}y_ky_j)_{,i} = 2\tau_{ij}y_j$$

is perpendicular to the surface given by Eq. (III–8). This means that the stress vector on A has a direction perpendicular to the tangent plane of the quadric surface at that point where the extension of n_i cuts the surface. The direction and normal component of the stress vector on A determine it completely, so by this method the quadric surface allows one to visualize readily the stress vectors associated with various area elements at P.

3. The most useful stress-state representation is given by the Mohr circle. Let the stress state in the local principal-axis system at P be τ_1, τ_2, τ_3; denote the principal axes by x_1, x_2, x_3. Consider now those area elements at P whose unit normals make an angle α with x_1. If N and S are the magnitudes of the normal and shear stresses on these elements, then

$$S^2 + N^2 = \tau_1^2 n_1^2 + \tau_2^2 n_2^2 + \tau_3^2 n_3^2,$$

which may be rewritten

$$\left(N - \frac{\tau_2 + \tau_3}{2}\right)^2 + S^2 = (\cos^2 \alpha)(\tau_1 - \tau_2)(\tau_1 - \tau_3) + \left(\frac{\tau_2 - \tau_3}{2}\right)^2. \quad \text{(III–9)}$$

Here $N = \tau_{ij} n_i n_j$ and so is positive if tensile, negative if compressive. Equation (III–9) implies that N and S lie on a circle with center at $\frac{1}{2}(\tau_2 + \tau_3)$ and with a radius which may be obtained graphically as shown in Fig. III–5. The values of N and S for area elements for which

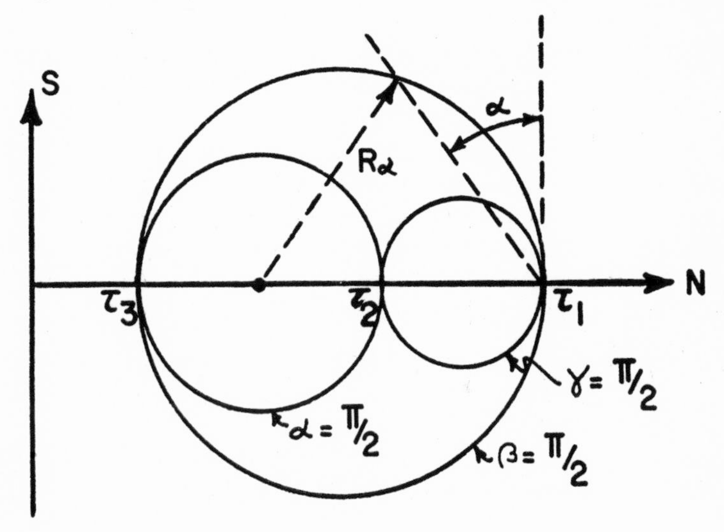

FIG. III–5. Mohr circle diagram.

$n_2 = \cos\beta$ or $n_3 = \cos\gamma$ also lie on circles, with centers at $\frac{1}{2}(\tau_1 + \tau_3)$ and $\frac{1}{2}(\tau_1 + \tau_2)$, respectively. Their radii are obtained similarly. The general situation is shown in Fig. III–6. Note that α is laid off from the τ_1-point, β from the τ_2-point, and γ from the τ_3-point.

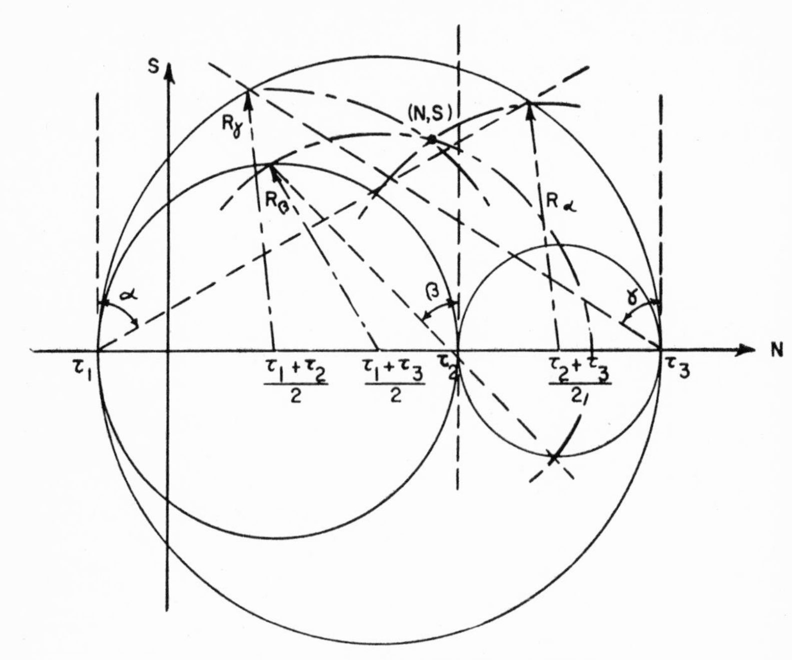

FIG. III–6. Complete Mohr circle diagram.

A little thought will show that (N, S) must always lie in the region between the solid circles. It follows that the maximum shear stress is one-half the largest difference between any two of the principal stresses, and occurs on an area element whose unit normal makes an angle of 45° with each of the corresponding principal axes. The quantities

$$s_1 = \tfrac{1}{2} \mid \tau_2 - \tau_3 \mid, \quad s_2 = \tfrac{1}{2} \mid \tau_1 - \tau_3 \mid, \quad s_3 = \tfrac{1}{2} \mid \tau_1 - \tau_2 \mid$$

are called *principal shears*. Note that

$$s_1^2 + s_2^2 + s_3^2 = \tfrac{1}{4}(3P_2 - P_1^2)$$

(cf. Sec. III–4, 3) and also that the s_i are unaltered by the superposition of a hydrostatic tension.

The last statement of Sec. III–4, 2 may now be used to show that the stress state equal to the sum of two stress states has a maximum shear stress which is not greater than the sum of the individual maximum shears.

An important special case of the Mohr circle arises if $\gamma = 90°$. The representative point then lies on a circle — called the dyadic circle — with center at $\frac{1}{2}(\tau_1 + \tau_2)$ and of radius $\frac{1}{2}|\tau_2 - \tau_1|$. The position of the point may be determined by laying off either of the angles α, β as shown in Fig. III–7. When the area element is perpendicular to the x_1-axis, the

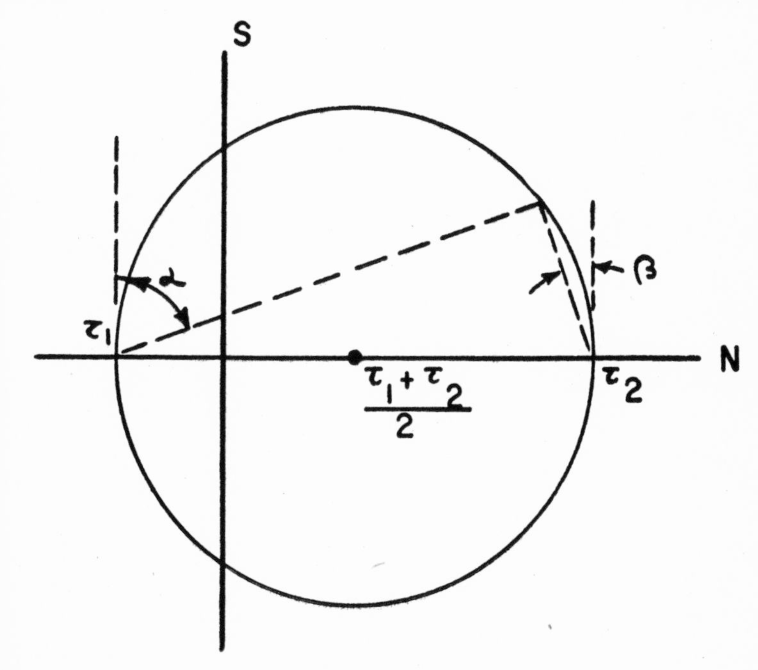

FIG. III–7. Dyadic circle.

corresponding representative point is at $(N = \tau_1, S = 0)$; as the area element rotates, the point moves around the circle at twice that rate of rotation. Whenever the point crosses the N-axis, the relative orientation of S reverses.

Since the center of the dyadic circle must lie on the N-axis, a knowledge of any pair of (N, S) values allows the circle to be drawn (the signs of S are immaterial). Alternatively, a knowledge of N for three directions is sufficient. Let the angles made by these three directions with the x_1 principal axis be α_1, α_2, α_3; since the angular separations are presumed known, the known quantities are $\alpha_2 - \alpha_1$ and $\alpha_3 - \alpha_2$. Draw three lines parallel to the S-axis at distances N_1, N_2, N_3 (Fig. III–8). Choose any

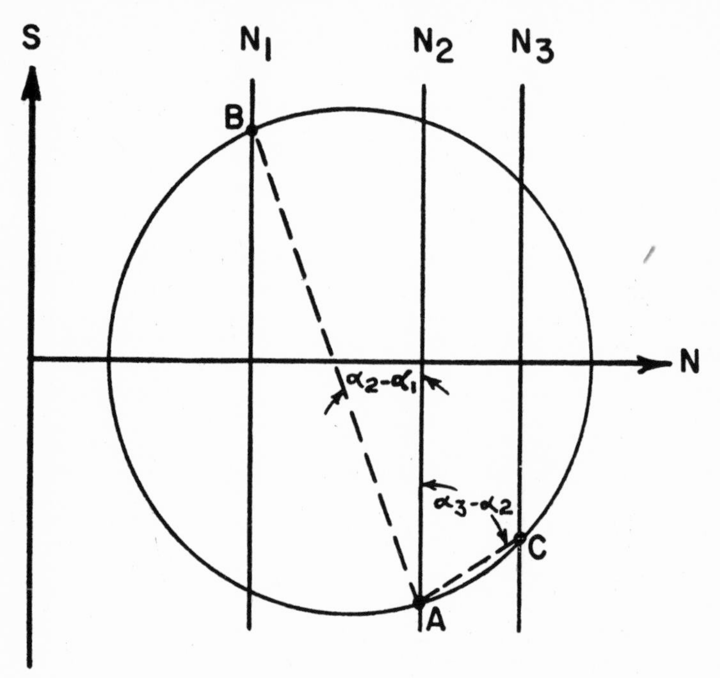

FIG. III–8. Geometric construction of dyadic circle.

point A on the N_2-line and lay off from it the angles $(\alpha_2 - \alpha_1)$ and $(\alpha_3 - \alpha_2)$, on opposite sides. In this way points B and C are found, which together with A determine the circle; the N-axis is drawn to pass through its center. That this method of construction produces the desired result follows from the fact that the angle intercepted by an arc at the center of a circle is twice the angle intercepted by the same arc at the circumference. In constructing this diagram, the relative magnitudes of N_1, N_2, N_3 are unimportant; however, the angle $(\alpha_2 - \alpha_1)$ is always laid off to the left of A, and the line projected in any direction to meet the N_1-line at B, and similarly $(\alpha_3 - \alpha_2)$ is always laid off to the right of A.

IV

Deformation

IV–1. Lagrangian and Eulerian Variables

Let X_i be a *fixed* Cartesian coordinate system with respect to which the motion of a material body may be described. Let the coordinates of a particular material particle (that is, a small cluster of molecules) at time $t = 0$ be a_i, and let the coordinates of the same material particle at time t be x_i. Then the displacement vector u_i is given by

$$u_i = x_i - a_i.$$

For example, in the two-dimensional motion given by

$$\begin{aligned} x_1 &= \tfrac{1}{2}(a_1 + a_2)e^t + \tfrac{1}{2}(a_1 - a_2)e^{-t}, \\ x_2 &= \tfrac{1}{2}(a_1 + a_2)e^t - \tfrac{1}{2}(a_1 - a_2)e^{-t}, \end{aligned} \qquad \text{(IV–1)}$$

the displacement vector is

$$\begin{aligned} u_1 &= \tfrac{1}{2}(a_1 + a_2)e^t + \tfrac{1}{2}(a_1 - a_2)e^{-t} - a_1, \\ u_2 &= \tfrac{1}{2}(a_1 + a_2)e^t - \tfrac{1}{2}(a_1 - a_2)e^{-t} - a_2. \end{aligned} \qquad \text{(IV–2)}$$

Here the displacements at time t are given in terms of the position (a_1, a_2) occupied by the particle at time $t = 0$. Alternatively, the same displacements could be specified in terms of the position (x_1, x_2) occupied by the particle at time t:

$$\begin{aligned} u_1 &= x_1 - \left[\tfrac{1}{2}(x_1 + x_2)e^{-t} + \tfrac{1}{2}(x_1 - x_2)e^t\right], \\ u_2 &= x_2 - \left[\tfrac{1}{2}(x_1 + x_2)e^{-t} - \tfrac{1}{2}(x_1 - x_2)e^t\right]. \end{aligned} \qquad \text{(IV–3)}$$

Equations (IV–2) and (IV–3) are examples of the use of *Lagrangian* and *Eulerian* variables, respectively. In the Lagrangian system, all quantities are expressed in terms of the initial position coordinates of each particle and time; in the Eulerian system, the independent variables are x_i and t, where x_i are the position coordinates at the time of interest.

Consider now any quantity ϕ, such as temperature, second component of displacement vector, third component of body force, and so forth. As a particular particle moves, it will observe a change in ϕ; the rate at which an observer attached to the particle would see ϕ alter is called the *material derivative* of ϕ and is denoted by $d\phi/dt$. If ϕ is expressed as $\phi(a_1, a_2, a_3, t)$, then

$$\frac{d\phi}{dt} = \frac{\partial\phi}{\partial t}, \tag{IV–4}$$

where all a_i are fixed in the partial differentiation, corresponding to the fact that we restrict our attention to one particular material particle and hence to one particular set of a_i. If on the other hand ϕ is expressed as $\phi(x_1, x_2, x_3, t)$, then all variables[1] will alter as the material particle moves, so that

$$\frac{d\phi}{dt} = \frac{\partial\phi}{\partial t} + \frac{\partial\phi}{\partial x_j}\frac{dx_j}{dt}. \tag{IV–5}$$

The expression $\partial\phi/\partial t$ in Eq. (IV–5) is to be calculated for fixed x_i and so is quite different from the $\partial\phi/\partial t$ used in Eq. (IV–4).

Clearly, the velocity of particle motion is the material derivative of position, so that

$$v_i = \frac{dx_i}{dt} = \frac{du_i}{dt}.$$

This implies that Eq. (IV–5) can be rewritten as

$$\frac{d\phi}{dt} = \frac{\partial\phi}{\partial t} + \frac{\partial\phi}{\partial x_j}v_j.$$

In the example above, the velocity vector is

$$v_1 = \frac{dx_1}{dt} = x_2,$$

$$v_2 = \frac{dx_2}{dt} = x_1.$$

The acceleration vector could be found by

$$a_i = \frac{d^2x_i}{dt^2} = \frac{dv_i}{dt}.$$

A good example of the distinction between Lagrangian and Eulerian variables is to be found in the equation of continuity. Let ρ be the den-

[1] The functional forms of $\phi(a_i, t)$ and $\phi(x_i, t)$ will of course be different, since ϕ depends in a different way upon the x_i than upon the a_i. We will, however, for brevity use the symbol ϕ in both cases and leave it to the context to determine which set of independent variables is contemplated.

sity of a medium, expressed first in terms of x_i and t. Then consider a closed surface S *fixed in space* and equate the rate of diminution of material inside the enclosed volume V, that is,

$$-\int \frac{\partial \rho}{\partial t}\, dV,$$

to the rate of material efflux from S, that is,

$$\int \rho v_i n_i \, dS = \int (\rho v_i)_{,i} \, dV,$$

by use of the divergence theorem of Gauss (Sec. I–13). Here, the partial differentiation is with respect to x_i. Since this must hold for any such volume, the integrands must be equal (cf. Sec. III–2) so that

$$\frac{\partial \rho}{\partial t} + (\rho v_i)_{,i} = 0.$$

or, equivalently,

$$\frac{d\rho}{dt} + \rho v_{i,i} = 0. \qquad \text{(IV–6)}$$

This is the equation of continuity in Eulerian form.

Alternatively, let ρ be expressed in terms of a_i and t, and consider a collection of material particles which originally occupies a volume V_0 and which at time t occupies a volume V. Since the mass within V does not alter,

$$\int \rho(a_i, t)\, dV = \int \rho(a_i, 0)\, dV_0.$$

If J denotes the Jacobian (Sec. I–11) of x_i with respect to a_i at time t, this equation becomes

$$\int \rho(a_i, t) J\, dV_0 = \int \rho(a_i, 0)\, dV_0,$$

and, since this must hold for any V_0,

$$\rho(a_i, t) J = \rho(a_i, 0) = \text{constant in time},$$

so that the equation of continuity becomes

$$\frac{d}{dt}(\rho J) = 0 \qquad \text{(IV–7)}$$

in Lagrangian form. The reader will find it useful to derive Eqs. (IV–6) and (IV–7) directly from one another.

It is occasionally necessary to evaluate an expression of the form

$$\frac{d}{dt} \int \phi \, dV$$

where the volume of integration contains always the same material particles and the expression is to be interpreted as the rate of change of the integral as seen by an observer moving with the volume. The differential operator cannot be brought under the integral sign, for the shape of the volume of integration is continually changing and this must be accounted for. However, if the expression is rewritten

$$\frac{d}{dt}\int \left(\frac{\phi}{\rho}\right) \rho\, dV,$$

then $\rho\, dV$ may be thought of as an element of mass, which is invariant, so that the region of integration no longer alters. Consequently,

$$\frac{d}{dt}\int \phi\, dV = \int \frac{d}{dt}\left(\frac{\phi}{\rho}\right) \rho\, dV. \tag{IV–8}$$

An alternative way of deriving Eq. (IV–8) is to transform to the volume V_0 and use Eq. (IV–7).

Finally, it may be noted that there is an interesting alternative interpretation of each of the coordinate sets (a_i), (x_i):

1. At time t, the material body is in a certain configuration. With each material particle in its final position, associate in imagination its original position coordinates a_i. (The surfaces of constant a_1, a_2, or a_3 will of course no longer be planes.) The (a_i) may then be thought of as providing a curvilinear coordinate net, related to the Cartesian coordinates (x_i) by $a_i = a_i(x_1, x_2, x_3, t)$.

2. Conversely, consider the body in its initial configuration and associate with each material particle the x_i coordinates that the particle will occupy at time t. Then this x_i-system provides a curvilinear coordinate net related to the Cartesian coordinates (a_i) by $x_i = x_i(a_1, a_2, a_3, t)$.

In the theory of elasticity, it is often only the initial and final configurations of the medium that are of interest, so that it is then not necessary to consider the dependence of various quantities on time. However, there are certain situations where it is convenient to discuss the time history of the deformation, and for that reason the general case has been considered in this section.

IV–2. Strain Tensor

As a deformable medium moves, the various portions of the medium will translate, rotate, and deform. The easiest way to distinguish between the deformation and the local rigid-body motion is to consider the change in distance between neighboring material particles. Suppose that two material particles, before the motion, have coordinates (a_i), $(a_i + da_i)$; after the motion, let their respective coordinates be (x_i), $(x_i + dx_i)$. Then the initial and final distances between these neighboring particles are given by

$$ds_0^2 = da_i\, da_i,$$
$$ds^2 = dx_i\, dx_i.$$

Only in the event of deformation is ds^2 different from ds_0^2. If the displacement vector is defined by

$$u_i = x_i - a_i,$$

the difference between ds^2 and ds_0^2 (which is easier to handle algebraically than $ds - ds_0$) is

$$\begin{aligned} ds^2 - ds_0^2 &= \left[\frac{\partial x_t}{\partial a_i}\frac{\partial x_t}{\partial a_j} - \delta_{ij}\right] da_i\, da_j \\ &= \left[\frac{\partial}{\partial a_i}(a_t + u_t)\cdot\frac{\partial}{\partial a_j}(a_t + u_t) - \delta_{ij}\right] da_i\, da_j \\ &= \left[\left(\delta_{it} + \frac{\partial u_t}{\partial a_i}\right)\left(\delta_{jt} + \frac{\partial u_t}{\partial a_j}\right) - \delta_{ij}\right] da_i\, da_j \\ &= \left[\frac{\partial u_i}{\partial a_j} + \frac{\partial u_j}{\partial a_i} + \frac{\partial u_t}{\partial a_i}\frac{\partial u_t}{\partial a_j}\right] da_i\, da_j. \end{aligned}$$

Define now

$$\eta_{ij} = \frac{1}{2}\left(\frac{\partial u_i}{\partial a_j} + \frac{\partial u_j}{\partial a_i} + \frac{\partial u_t}{\partial a_i}\frac{\partial u_t}{\partial a_j}\right), \qquad \text{(IV–9)}$$

so that

$$ds^2 - ds_0^2 = 2\eta_{ij}\, da_i\, da_j. \qquad \text{(IV–10)}$$

The set of quantities (η_{ij}) is called the *Lagrangian strain tensor*. The factor $\frac{1}{2}$ is inserted in its definition for convenience of subsequent physical interpretation. Similarly, the x_i variables could have been used in the calculation, which would then have given

$$ds^2 - ds_0^2 = 2e_{ij}\, dx_i\, dx_j, \qquad \text{(IV–11)}$$

where

$$e_{ij} = \frac{1}{2}\left(\frac{\partial u_i}{\partial x_j} + \frac{\partial u_j}{\partial x_i} - \frac{\partial u_t}{\partial x_i}\frac{\partial u_t}{\partial x_j}\right). \qquad \text{(IV–12)}$$

The set of quantities (e_{ij}) is called the *Eulerian strain tensor*.

It may be noted that there is a difference in form between Eqs. (IV–9) and (IV–12), in that a + sign in front of the product term in Eq. (IV–9) corresponds to a − sign in Eq. (IV–12). Each of η_{ij}, e_{ij} is symmetric in its two subscripts. A necessary and sufficient condition that there be no deformation is that each of η_{ij}, e_{ij} vanish.

Before discussing the tensor character of η_{ij}, e_{ij}, it is worth while to examine the physical meaning of these quantities. A simple physical interpretation exists only in the case of small deformation; for large deformation, Eqs. (IV–10) and (IV–11) cannot be simplified. Suppose then that the deformation (although not necessarily the displacement u_i) is small, so that $ds_0 \cong ds$. Then for the case of a line element which after the motion is parallel to the X_1-axis, Eq. (IV–11) may be written

$$(ds - ds_0)(ds + ds_0) = 2e_{11}\, dx_1\, dx_1 = 2e_{11}\, ds^2,$$

that is,

$$\frac{ds - ds_0}{ds_0} = e_{11}\left[\frac{ds}{ds_0} \cdot \frac{2\, ds}{ds + ds_0}\right]$$

and, since the smallness of the deformation implies that the quantity in brackets is approximately unity, it follows that e_{11} may be interpreted as the fractional elongation of a line element which *after* the motion is parallel to the X_1-axis. Similarly, e_{22} and e_{33} represent the fractional elongations of line elements which after the motion are parallel to the X_2-, X_3-axes respectively. A similar calculation for the Lagrangian strain shows that η_{11}, η_{22}, η_{33} represent the fractional elongations of fibers which *before* the motion are parallel to the X_1-, X_2-, X_3-axes respectively.

Continuing this physical interpretation study for small deformations, consider now the typical off-diagonal term e_{12}. Let $dx_i^{(1)}$ and $dx_i^{(2)}$ represent two line elements (that is, material fibers), emanating from the same point, which after the motion are respectively parallel to the X_1- and X_2-axes. Denote the initial angular separation between the fibers by $\frac{1}{2}\pi + \theta$, so that

$$\cos\left(\tfrac{1}{2}\pi + \theta\right) = \frac{da_i^{(1)}\, da_i^{(2)}}{|\, da_i^{(1)}\, | \cdot |\, da_i^{(1)}\, |}$$

$$= \frac{(dx_i^{(1)} - u_{i,k}\, dx_k^{(1)})(dx_i^{(2)} - u_{i,t}\, dx_t^{(2)})}{|\, da_i^{(1)}\, | \cdot |\, da_i^{(2)}\, |}$$

$$= \frac{-\,2e_{ik}\, dx_i^{(1)}\, dx_k^{(2)}}{|\, da_i^{(1)}\, | \cdot |\, da_i^{(2)}\, |}$$

$$\cong \frac{-\,2e_{12}}{(1 - e_{11})(1 - e_{22})} \cong -\,2e_{12}.$$

But $\cos\left(\frac{1}{2}\pi + \theta\right) \cong -\theta$, so that e_{12} is one-half the decrease in angle between two line elements which are parallel to the X_1-, X_2-axes after the motion. Similarly, η_{12} would represent one-half the decrease in angle between two line elements which are parallel to the X_1-, X_2-axes before the motion. The other off-diagonal components of strain are interpreted similarly. Consequently, the off-diagonal terms represent shear.

In the case of small deformation (again, nowhere has it been assumed that the u_i are small), the various components of strain are therefore closely linked to elementary ideas of strain. In the older literature, the fractional elongations and changes in angle (these latter not multiplied by $\frac{1}{2}$) are defined as the strains, but this has the disadvantage that the resulting set of quantities is not a tensor and so is not subject to the simple rules of manipulation shortly to be discussed; further, separate equations would have to be written for elongation and shear terms, and in general considerations this is quite awkward.

If the displacement derivatives are small compared to unity, then

$$
\begin{aligned}
e_{ij} &\cong \frac{1}{2}\left(\frac{\partial u_i}{\partial x_j} + \frac{\partial u_j}{\partial x_i}\right), \\
\eta_{ij} &\cong \frac{1}{2}\left(\frac{\partial u_i}{\partial a_j} + \frac{\partial u_j}{\partial a_i}\right).
\end{aligned}
\qquad \text{(IV–13)}
$$

The physical interest of this approximation may be exemplified by the fact that the elastic limit of a steel rod in tension will be exceeded for fractional elongations greater than about 0.1 percent. The advantage of the approximation lies in the fact that the equations between strain and displacement are then linear, a traditional advantage in analysis. For small displacements, there is little difference between x_i and a_i, so that an additional result is that

$$e_{ij} \cong \eta_{ij}.$$

Physical visualization of the meaning of the partial derivatives in Eq. (IV–13) leads to the same interpretation in terms of fractional elongations and angular changes previously obtained more generally. There are, however, situations where the linearity approximation may not be valid — usually when one over-all dimension of an elastic body is of a different order of magnitude than the others, as in the case of a long beam. Despite the smallness of the deformation in such cases, the displacements and displacement derivatives may be large (corresponding to large rotation) and it is the size of these latter quantities that controls the validity of the linearity approximation. Another situation in which the nonlinearity terms cannot be neglected arises in those stability problems in which all linear terms cancel.

It may be worth while to emphasize the occasional importance of the nonlinear terms by calculating the strains for the following combination of magnification and rotation:

$$u_1 = A(a_1 \cos \alpha - a_2 \sin \alpha) - a_1,$$
$$u_2 = A(a_2 \cos \alpha + a_1 \sin \alpha) - a_2,$$
$$u_3 = 0.$$

Then the exact values for η_{ij} are

$$\eta_{11} = \eta_{22} = \tfrac{1}{2}(A^2 - 1),$$
$$\eta_{12} = \eta_{21} = 0.$$

However, neglect of the product terms would give

$$\eta_{11}' = \eta_{22}' = A \cos \alpha - 1,$$
$$\eta_{12}' = \eta_{21}' = 0,$$

so that, even if the deformation is small ($A \cong 1$), the values for η_{11}' and η_{22}' will be highly inaccurate unless α is also small.

Returning to the general nonlinear case, the following easily verified formulas may be collected for reference (du_i/dt is denoted by $\dot{u}_i$, $d\eta_{ij}/dt$ by $\dot{\eta}_{ij}$, and so on):

$$\eta_{ij} = \frac{1}{2}\left(\frac{\partial x_s}{\partial a_i}\frac{\partial x_s}{\partial a_j} - \delta_{ij}\right), \tag{IV–14}$$

$$e_{ij} = \frac{1}{2}\left(\delta_{ij} - \frac{\partial a_s}{\partial x_i}\frac{\partial a_s}{\partial x_j}\right), \tag{IV–15}$$

$$\eta_{ij} = e_{st}\frac{\partial x_s}{\partial a_i}\frac{\partial x_t}{\partial a_j}, \tag{IV–16}$$

$$e_{ij} = \eta_{st}\frac{\partial a_s}{\partial x_i}\frac{\partial a_t}{\partial x_j}, \tag{IV–17}$$

$$\frac{d}{dt}\left(\frac{\partial x_r}{\partial a_k}\right) = \frac{\partial \dot{u}_r}{\partial a_k}, \tag{IV–18}$$

$$\frac{d}{dt}\left(\frac{\partial a_r}{\partial x_k}\right) = -\frac{\partial a_r}{\partial x_s}\frac{\partial \dot{u}_s}{\partial x_k}, \tag{IV–19}$$

$$\dot{\eta}_{ij} = \frac{1}{2}\left(\frac{\partial \dot{u}_s}{\partial x_t} + \frac{\partial \dot{u}_t}{\partial x_s}\right)\frac{\partial x_s}{\partial a_i}\frac{\partial x_t}{\partial a_j}, \tag{IV–20}$$

$$\dot{e}_{ij} = \frac{1}{2}\left(\frac{\partial \dot{u}_i}{\partial x_j} + \frac{\partial \dot{u}_j}{\partial x_i}\right) - e_{im}\frac{\partial \dot{u}_m}{\partial x_j} - e_{jm}\frac{\partial \dot{u}_m}{\partial x_i}. \tag{IV–21}$$

Note that $\partial x_i/\partial a_j$, $\partial a_i/\partial x_j$ are the elements of the Jacobian or inverse Jacobian of the functional relation between x_i and a_i. Note also that Eq. (IV–20) implies that a necessary and sufficient condition that a body be moving without deformation is that

$$\frac{\partial \dot{u}_i}{\partial x_j} + \frac{\partial \dot{u}_j}{\partial x_i} = 0.$$

Finally, it may be noticed that the same kind of term is added in Eq. (IV–9) as is subtracted in Eq. (IV–12); this leads one to ask whether a coordinate system "halfway between" the a_i-and x_i-systems might not be of interest. Define

$$b_i = a_i + \tfrac{1}{2}u_i$$

and consider the curvilinear coordinate system b_i as basic. Then calculation gives

$$ds^2 - ds_0^2 = 2\beta_{ij}\, db_i\, db_j,$$

where

$$\beta_{ij} = \frac{1}{2}\left(\frac{\partial u_i}{\partial b_j} + \frac{\partial u_j}{\partial b_i}\right).$$

There is thus both a gain and a loss in using this coordinate system — a gain because of the linearity of the exact strain components, but a loss because of the non-Cartesian nature of the coordinate system.

IV–3. Rotation Tensor

Consider again two neighboring material particles initially separated by the vector distance da_i and finally by the vector distance dx_i. Using first Lagrangian variables,

$$\begin{aligned} dx_i &= da_i + \frac{\partial u_i}{\partial a_j}\, da_j \\ &= \left[\delta_{ij} + \frac{1}{2}\left(\frac{\partial u_i}{\partial a_j} + \frac{\partial u_j}{\partial a_i} + \frac{\partial u_s}{\partial a_i}\frac{\partial u_s}{\partial a_j}\right) + \frac{1}{2}\left(\frac{\partial u_i}{\partial a_j} - \frac{\partial u_j}{\partial a_i} - \frac{\partial u_s}{\partial a_i}\frac{\partial u_s}{\partial a_j}\right)\right] da_j \\ &= [\delta_{ij} + \eta_{ij} + \gamma_{ij}]\, da_j, \end{aligned} \qquad \text{(IV–22)}$$

where γ_{ij} is defined by this equation.

The set of quantities γ_{ij} is called the *Lagrangian rotation tensor*. The genesis of its name is that if the body is moving without deformation, so that $\eta_{ij} = 0$, then in Eq. (IV–22) the contribution of γ_{ij} can represent rotation alone (translation being removed by the partial differentiation of the u_i). Similarly, the *Eulerian rotation tensor* is defined by

$$\omega_{ij} = \frac{1}{2}\left(\frac{\partial u_i}{\partial x_j} - \frac{\partial u_j}{\partial x_i} + \frac{\partial u_s}{\partial x_i}\frac{\partial u_s}{\partial x_j}\right),$$

with

$$da_i = [\delta_{ij} + e_{ij} + \omega_{ij}]\, dx_j. \tag{IV-23}$$

If the deformation η_{ij} is zero (or in any event very small), then γ_{ij} in Eq. (IV–22) may be thought of as an operator which must be applied to a fiber da_i in order to rotate this fiber around into its new position. If the deformation is not small, there is no easy way of physically separating the motion into a pure deformation and a pure rotation, for the two kinds of motion are intimately associated with one another. One way of trying to make such a separation would be to define the rotation at a point as the average (in some sense) of the rotations of all fibers emanating from this point; however, the results become too complicated for interest.[2] Another way of defining rotation is to use the fact (to be proved in the next section) that the strain tensor η_{ij} has principal axes at the point in question. These three axes are mutually orthogonal before and after the motion. A spherical shell of matter surrounding the point becomes distorted into an ellipsoid whose axes coincide with the principal axes; this ellipsoid then rotates. A very natural way to define rotation at the point would then be in terms of the amount of rotation undergone by this ellipsoid. Unfortunately, the results are again complicated. Consequently, it seems profitless in general to search for physical interpretations beyond those directly implied by Eqs. (IV–22) and (IV–23).

If the displacement derivatives are small, then only the linear terms need be used in the definitions of γ_{ij} and ω_{ij}. For example,

$$\gamma_{12} = \frac{1}{2}\left(\frac{\partial u_1}{\partial a_2} - \frac{\partial u_2}{\partial a_1}\right),$$

and this now has a clear-cut interpretation in terms of rotations of fibers initially parallel to the X_1-, X_2-axes around the X_3-axis.

[2] Novozhilov, *Foundations of the nonlinear theory of elasticity* (Graylock Press, Rochester, New York, 1953), p. 27, gives a modified analysis along these lines. Rotations are defined in terms of angles between the initial line elements and their final projections onto the coordinate planes; the *tangents* of these angles are averaged for all fibers. This (vector) average is found to be proportional to the linear part of the off-diagonal terms in γ_{ij}; the coefficients are however difficult to interpret, as they involve the linear parts of η_{ij}.

In the general case, the various displacement derivatives are, by the derivation of Eq. (IV–22), expressible in terms of the strain and rotation tensors. In the Lagrangian system,

$$\frac{\partial u_i}{\partial a_j} = \eta_{ij} + \gamma_{ij},$$

so that

$$\eta_{11} = \frac{\partial u_1}{\partial a_1} + \frac{1}{2}[(\eta_{11} + \gamma_{11})^2 + (\eta_{21} + \gamma_{21})^2 + (\eta_{31} + \gamma_{31})^2],$$

and so on. Such formulas are occasionally useful in correcting the linear approximation to the η_{ij}. For example, if the displacement derivatives were small, with rotation more important than strain, then the bracketed expression might be approximated by

$$\tfrac{1}{2}[\gamma_{21}^2 + \gamma_{31}^2].$$

IV–4. Properties of Strain

Consider first the body in its initial configuration. At each point a_i, the quantities η_{ij} and γ_{ij} are defined in terms of derivatives of the vector u_i, and so (by Chapter II) must be tensors with respect to rotations of the X_i-system. If, for example, the rotation $X_i' = \alpha_{ij}X_j$ is used, so that the before-deformation coordinates of a given material point change from a_i to a_i', then

$$\eta_{ij}' = \alpha_{is}\alpha_{jt}\eta_{st},$$
$$\gamma_{ij}' = \alpha_{is}\alpha_{jt}\gamma_{st}.$$

Omitting the nonlinear terms in η_{ij} and γ_{ij} would not alter their tensor character.

If, similarly, the final configuration of the body is considered, the quantities e_{ij} and ω_{ij} (with or without their nonlinear terms) are tensors with respect to rotation of the X_i-system. For the rotation in question,

$$e_{ij}' = \alpha_{is}\alpha_{jt}e_{st},$$
$$\omega_{ij}' = \alpha_{is}\alpha_{jt}\omega_{st}.$$

Consider a unit vector n_i, as specified in the fixed X_i-system. Define *strain vectors*

$$e_i = e_{ij}n_j, \qquad \eta_i = \eta_{ij}n_j, \tag{IV–24}$$

and examine, for example, the physical interpretation of e_i in the case of small deformations. The easiest way of doing this is to use a local coordinate system whose X_1'-axis coincides with n_i, so that $n_i' = (1, 0, 0)$, and whose X_2'-axis lies in the common plane of e_i and n_i. Then since the tensor equation (IV–24) holds also in this coordinate system,

$$\begin{aligned} e_i' &= e_{ij}'n_j' \\ &= (e_{11}', e_{21}', e_{31}'). \end{aligned}$$

Now $e_3' = 0$, so that $e_{31}' = 0$. The remaining two components have the following physical interpretation:

1. e_1' = fractional elongation of a fiber which after the motion lies along the X_1'-axis.

2. e_2' = change in angle between two fibers which after the motion lie along the X_1'-, X_2'-axes. (There is no change in angle for the X_1'-, X_3'-axes.)

Returning to the original coordinate system, we note that e_1' is the component of e_i in the direction of n_i and that e_2' is the component of e_i perpendicular to n_i.

The strain vector η_i is interpreted analogously with respect to positions of fibers before the motion.

Again, this physical interpretation ceases to be valid when the deformation is large. However, whether the deformation is large or small, it is possible to ask whether there exist directions n_i for which e_i is parallel to n_i. For such a direction, the quantities e_{31}' and e_{32}' would be zero, which by Sec. IV–2 implies that fibers which after the motion are perpendicular to the fiber then occupying n_i were also (exactly) perpendicular to the same material fiber before the motion.

We are therefore led to the equations $e_{ij}n_j = \lambda n_i$; because of the symmetry of e_{ij}, the existence of at least three such directions, mutually perpendicular, may be inferred exactly as in Sec. III–3. Thus there exists a principal-axis system, in which all off-diagonal e_{ij} vanish. The material fibers occupying these directions after the motion were also mutually perpendicular before the motion. The three λ-roots of the determinantal equation $| e_{ij} - \lambda\delta_{ij} | = 0$ are the values of the nonvanishing components of the strain tensor in a coordinate system lined up with the principal axes. The maximum and minimum values of e_in_i will be found among those n_i which have the direction of a principal axis (cf. Sec. III–4); since

$$2e_in_i = 2e_{ij}n_in_j = 2e_{ij}\frac{dx_i}{ds}\frac{dx_j}{ds} = 1 - \left(\frac{ds_0}{ds}\right)^2,$$

these are also the directions of greatest and least fiber extension.

The situation with respect to invariants is the same as in Sec. (III–4). Define

$$E_1 = e_{ii},$$
$$E_2 = \text{sum of principal two-rowed minors of } (e_{ij}), \qquad \text{(IV–25)}$$
$$E_3 = \text{determinant of } (e_{ij}).$$

As before, these quantities do not depend on the coordinate system used. A further property of interest is obtained by noting that from Eq. (IV–7) the ratio of final to initial density is given by

$$\frac{\rho}{\rho_0} = \left|\frac{\partial a_i}{\partial x_j}\right|,$$

where the right-hand side is the Jacobian of the a_i with respect to the x_i. It follows that

$$\frac{\rho}{\rho_0} = \left(\left|\frac{\partial a_s}{\partial x_i}\frac{\partial a_s}{\partial x_j}\right|\right)^{\frac{1}{2}}$$
$$= (|\ \delta_{ij} - 2e_{ij}\ |)^{\frac{1}{2}}$$
$$= (1 - 2E_1 + 4E_2 - 8E_3)^{\frac{1}{2}}. \qquad \text{(IV–26)}$$

The remainder of the discussion in Secs. III–4 and III–5 may also be carried over to the case of the strain tensor. Thus the condition for a state of pure shear is the same, the projection theorems and their consequences are identical, and similar geometric interpretations may be given. The continuity-type argument may be used here also, to show that at each point there exists at least one fiber whose direction is not altered by the motion. As a matter of fact, all of these properties are simply general properties of second-order symmetric tensors.

If the linearized form of e_{ij} had been used, these properties would again hold. Finally, if we had chosen to discuss η_{ij} instead of e_{ij}, similar properties would result. It might be expected that those fibers which before the motion coincide with the principal axes of η_{ij} will after the motion coincide with the principal axes of e_{ij}, and this follows from the orthogonality of principal fibers before and after deformation; alternatively, if

$$e_{ij}n_j = \lambda n_i$$

for some direction n_i, then setting $n_j = dx_j/ds$ for this particular direction gives

$$e_{ij}\frac{\partial x_j}{\partial a_t}\frac{da_t}{ds} = \lambda\frac{\partial x_i}{\partial a_r}\frac{da_r}{ds}.$$

Multiply through by $(ds/ds_0)(\partial x_i/\partial a_p)$ and use Eq. (IV–16) to give

$$\eta_{pt}\frac{da_t}{ds_0} = \lambda \frac{\partial x_i}{\partial a_p}\frac{\partial x_i}{\partial a_r}\cdot\frac{da_r}{ds_0}.$$

Defining

$$n_t^0 = \frac{da_t}{ds_0}$$

(that is, n_t^0 represents a unit vector in the original direction of that fiber which after the motion coincides with n_i) and using Eq. (IV–14) gives finally

$$\eta_{pt}n_t^0 = \frac{\lambda}{1-2\lambda}n_p^0, \tag{IV–27}$$

which proves the statement. Since the coefficient of n_p^0 on the right-hand side is a principal value of η_{ij}, and λ is a principal value of e_{ij}, this result also provides the relation between the principal values of the two strain tensors. (The basic definition of e_{ij} ensures that λ cannot be $\geq \frac{1}{2}$.) This relation allows us to express the invariants N_1, N_2, N_3 of the η_{ij}, defined similarly to Eq. (IV–25), in terms of E_1, E_2, E_3:

$$N_1 = \frac{E_1 - 4E_2 + 12E_3}{1 - 2E_1 + 4E_2 - 8E_3}, \qquad E_1 = \frac{N_1 + 4N_2 + 12N_3}{1 + 2N_1 + 4N_2 + 8N_3},$$

$$N_2 = \frac{E_2 - 6E_3}{1 - 2E_1 + 4E_2 - 8E_3}, \qquad E_2 = \frac{N_2 + 6N_3}{1 + 2N_1 + 4N_2 + 8N_3},$$

$$N_3 = \frac{E_3}{1 - 2E_1 + 4E_2 - 8E_3}, \qquad E_3 = \frac{N_3}{1 + 2N_1 + 4N_2 + 8N_3}.$$

The density ratio is

$$\frac{\rho_0}{\rho} = (1 + 2N_1 + 4N_2 + 8N_3)^{\frac{1}{2}}. \tag{IV–28}$$

IV–5. Compatibility

1. Consider first the case of small displacement derivatives so that

$$e_{ij} = \tfrac{1}{2}(u_{i,j} + u_{j,i})$$

(where no distinction is made between Lagrangian and Eulerian variables). If one is given a set of displacements u_i, then a set of e_{ij} may certainly be calculated by these equations. However, the converse problem often arises in elasticity, namely, if one is given a set of e_{ij}, is it possible to calculate a corresponding set of displacements? In general, the answer is no; for example, it may be checked by direct substitution that, if such a set of u_i existed, then

$$2e_{12,12} = e_{11,22} + e_{22,11},$$

and the arbitrarily given e_{ij} need not satisfy this equation.

It turns out that there is a set of six equations which, if satisfied by the given e_{ij}, will ensure that a set of u_i does exist. These equations are called the *equations of compatibility*. To derive them, we begin with the given e_{ij} and try to calculate from them the u_i; it will be found that this calculation is possible only if certain conditions are satisfied.

Given any set of e_{ij}, the rigid-body part of the motion is still arbitrary; choose a point Q as reference point and let the displacement and rotation of this point be u_i^Q, ω_{ij}^Q. Then using the identity

$$\omega_{ij,k} = e_{ik,j} - e_{jk,i}, \tag{IV–29}$$

the rotation at any other point P will be given by

$$\begin{aligned} \omega_{ij}^p &= \omega_{ij}^q + \int_Q^P \omega_{ij,k}\, dx_k \\ &= \omega_{ij}^Q + \int_Q^P (e_{ik,j} - e_{jk,i})\, dx_k. \end{aligned} \tag{IV–30}$$

If a physically possible displacement actually does exist, the integral must be independent of the path. For any choice of values i, j, the integrand has the form $\int A_k\, dx_k$, and by Sec. I–13 the condition for this to be independent of the path is that $e_{psk}A_{k,s} = 0$. This condition becomes

$$e_{psk}(e_{ik,js} - e_{jk,is}) = 0; \tag{IV–31}$$

consequently Eq. (IV–31) is a necessary and sufficient condition that ω_{ij}^P can be calculated. Consider next u_i:

$$\begin{aligned} u_i^P &= u_i^Q + \int_Q^P u_{i,j}\, dx_j \\ &= u_i^Q + \int_Q^P (e_{ij} + \omega_{ij})\, dx_j \\ &= u_i^Q + \int_Q^P [e_{ij} + (\omega_{is}x_s)_{,j} - \omega_{is,j}x_s]\, dx_j \\ &= u_i^Q + [\omega_{is}x_s]_Q^P + \int_Q^P (e_{ij} - \omega_{is,j}x_s)\, dx_j. \end{aligned} \tag{IV–32}$$

If Eq. (IV–31) is satisfied, then $\omega_{is}{}^{P}$ exists, so that the quantity

$$[\omega_{is}x_s]_Q^P$$

is independent of the path; it is now also necessary that the integral in Eq. (IV–32) be independent of the path, and, by the same reasoning as that used in deriving Eq. (IV–31), the condition is found to be

$$e_{psk}x_j(e_{ik,js} - e_{jk,is}) = 0,$$

which is certainly satisfied if Eq. (IV–31) is. Consequently Eq. (IV–31) is the necessary and sufficient condition that a set of u_i exists. An equivalent form of Eq. (IV–31) is

$$e_{psk}e_{rij}e_{ik,js} = 0. \qquad \text{(IV–33)}$$

This equation is symmetric in p and r, so that there are only six different equations obtained by various choices of p and r. Thus the equations of compatibility are:

$$\begin{aligned} e_{11,23} &= (-e_{23,1} + e_{31,2} + e_{12,3})_{,1}, \\ e_{22,31} &= (-e_{31,2} + e_{12,3} + e_{23,1})_{,2}, \\ e_{33,12} &= (-e_{12,3} + e_{23,1} + e_{31,2})_{,3}, \\ 2e_{12,12} &= e_{11,22} + e_{22,11}, \\ 2e_{23,23} &= e_{22,33} + e_{33,22}, \\ 2e_{31,31} &= e_{33,11} + e_{11,33}. \end{aligned} \qquad \text{(IV–34)}$$

It may be noted that, if the first two of these equations are differentiated with respect to x_2, x_1 respectively and added, the result is identical with that obtained by differentiating the fourth equation with respect to x_3. Similar results hold for the other equations. In a sense, this means that only three of these six equations are strongly informative.

It is possible to obtain from Eq. (IV–33) another set of six equations which are equivalent to Eq. (IV–34) but which are for some purposes more useful. Equation (IV–33) may be written

$$e_{ik,js} - e_{is,jk} - e_{jk,is} + e_{js,ik} = 0.$$

Setting $s = j$ gives

$$e_{ik,jj} - e_{ij,jk} - e_{kj,ji} + e_{jj,ik} = 0. \qquad \text{(IV–35)}$$

Because of symmetry, only six are independent. Furthermore, the three equations of (IV–35) for which $i \neq k$ are at once identical with the first three of (IV–34), and the remaining three equations of (IV–35) are easily combined to yield the last three of (IV–34) (or vice versa); consequently the two sets (IV–34) and (IV–35) are completely equivalent. The form of Eqs. (IV–35) is such that where these equations are expressed in terms of stresses, by use of the stress-strain relations to be subsequently derived, some of the resulting combinations of stress derivatives may be simplified by use of the equilibrium equations.

The equations of compatibility (IV–33) were derived on the basis that the difference in displacement between two points, as expressed by an integral, should be independent of the path linking these points. Use was made of Stokes's theorem, and consequently it was tacitly assumed that any two paths linking these points could be deformed continuously into one another, that is, that there were no such holes in the region as to prevent this. If there are such holes (for example, if the body is in the form of a toroid), then the equations must certainly still hold in order to ensure that adjacent paths of integration give the same result; however, it is necessary to add the additional condition that integrations around opposite sides of the hole should yield the same result. If a "dislocation" has occurred (for instance, the toroid has been sheared along one radial plane), then this second condition must be modified appropriately.

2. Consider next the general case of large deformation. It may be expected that the general structure of the compatibility equation will be similar to Eq. (IV–33), with, however, some complications arising from nonlinearity. The problem would be a difficult one, were it not for the fact that appeal can be made to a well-known and powerful theorem of general tensor analysis.

Suppose first that the Eulerian quantities e_{ij} are given as specific functions of x_i, and it is desired to determine whether or not corresponding u_i exist — or, equivalently, whether or not corresponding a_i exist. Another way of formulating the question is to ask whether or not it is possible to assign curvilinear (Sec. IV–1) coordinates a_i to each point whose Cartesian coordinates are x_i, such that if the body were deformed so that the originally nonplanar surfaces of constant a_i became planes parallel to the coordinate axes, thus making the a_i (carried along with the material points) into Cartesian coordinates, then the distance ds_0 between adjacent points would be given by

$$ds_0^2 = da_i\, da_i = (\delta_{ij} - 2e_{ij})\, dx_i\, dx_j.$$

But the necessary and sufficient condition that a set of curvilinear coordinates a_i exist such that the quadratic form $(\delta_{ij} - 2e_{ij})\, dx_i\, dx_j$ becomes $da_i\, da_i$ is from general tensor analysis [3] that a certain set of six independent equations be satisfied, corresponding to the vanishing of the curvature tensor. The result is:

$$\frac{\partial^2 g_{rn}}{\partial x_s \partial x_m} + \frac{\partial^2 g_{sm}}{\partial x_r \partial x_n} - \frac{\partial^2 g_{rm}}{\partial x_s \partial x_n} - \frac{\partial^2 g_{sn}}{\partial x_r \partial x_m} + 2g^{pq}(\Gamma_{rn;p}\Gamma_{sm;q} - \Gamma_{rm;p}\Gamma_{sn;q}) = 0, \quad \text{(IV–36)}$$

where

$$g_{ij} = \delta_{ij} - 2e_{ij},$$

$$g = \text{determinant of the } g_{ij} \quad \text{(cf. Eq. (IV–26))},$$

$$g^{ij} = \frac{1}{g}[\text{cofactor of } g_{ij} \text{ in } g],$$

$$\Gamma_{ij;k} = \frac{1}{2}\left(\frac{\partial g_{ik}}{\partial x_j} + \frac{\partial g_{jk}}{\partial x_i} - \frac{\partial g_{ij}}{\partial x_k}\right).$$

Note the summation on indices p and q. The second part of the equation contains the nonlinear contributions; if they were omitted, Eq. (IV–33) would be obtained. Again, the six independent equations in Eq. (IV–36) are not strongly independent; their derivatives are related to one another by the Bianchi identity.[4]

In Lagrangian variables, the results are very similar. Write

$$ds^2 = (\delta_{ij} + 2\eta_{ij})da_i\, da_j$$

and the condition for existence of coordinates x_i such that $ds^2 = dx_i\, dx_i$ is again Eq. (IV–36), with now $g_{ij} = \delta_{ij} + 2\eta_{ij}$ and all partial differentiation carried out with respect to the a_i.

[3] See, for example, J. L. Synge and A. Schild, *Tensor calculus* (University of Toronto Press, Toronto, 1949), pp. 84, 106.

[4] Synge and Schild, p. 87.

V

Basic Equations of Linear Elasticity

V–1. Relation Between Stress and Strain

In linear elasticity, all displacements and displacement derivatives are assumed sufficiently small that the nonlinear terms in the strain and rotation tensors can be omitted and also that the distinction between Lagrangian and Eulerian variables becomes negligible.

An *elastic* body may be defined to be one for which the stress depends only on the deformation and not on the history of that deformation. If the relation is linear,

$$\tau_{ij} = c_{ijkm} e_{km}, \tag{V–1}$$

where the c_{ijkm} are constants, then the material is said to be linearly elastic. Equation (V–1) may be thought of as the first term in a power-series expansion of a more general functional relation (note that there can be no constant term in the expansion if zero strain is to imply zero stress); for a linearly elastic body, this first term is then, by definition, sufficiently exact for practical purposes. It may also be remarked that Eq. (V–1) is the simplest generalization of the linear dependence of stress on strain observed in the familiar Hooke's-law experiment in which a metal wire is stretched by a tensile load, and consequently Eq. (V–1) is often referred to as the *generalized Hooke's law.* Since deformation may result from thermal effects as well as stress, it is in general necessary to incorporate a temperature dependence in Eq. (V–1). For example, experiment shows that the elastic constants as measured under conditions of rapid and slow loading are slightly different. Thermal effects will be discussed in Chapter IX; for the present, they will be assumed unimportant.

In a crystalline body, the elastic properties are found to depend on the orientation of the body. However, many metals have a sufficiently small and random-oriented crystal grain structure that the elastic properties as averaged over a few grains are essentially independent of orientation, so that such metals may be considered *isotropic*. Even if forming processes such as rolling or drawing have made the material anisotropic, the degree of anisotropy is often small enough that practical answers can be obtained on the basis of assumed isotropy. Except in Sec. V–7, it will be assumed that the elastic body is isotropic.

In an isotropic body, τ_{12}, for example, must depend upon e_{32} in the same way irrespective of the coordinate system being used. This means that the numerical values of c_{1232} — and more generally of each c_{ijkm} — must be the same in all coordinate systems, so that c_{ijkm} is a fourth-order isotropic tensor, whose general form is given by (Sec. II–8)

$$c_{ijkm} = \alpha\delta_{ij}\delta_{km} + \beta\delta_{ik}\delta_{jm} + \gamma\delta_{im}\delta_{jk}.$$

Equation (V–1) becomes

$$\tau_{ij} = \alpha\delta_{ij}e_{kk} + \beta e_{ij} + \gamma e_{ji},$$

and because of the symmetry of e_{ij} this may be written

$$\tau_{ij} = 2\mu e_{ij} + \lambda\delta_{ij}e_{kk}, \qquad \text{(V–2)}$$

where conventional notation in terms of Lamé's constants μ and λ has been used. Setting $i = j$ gives a formula for e_{kk} in terms of τ_{kk}; if this result is substituted into Eq. (V–2), the equation may be solved for e_{ij} to give

$$2\mu e_{ij} = \tau_{ij} - \frac{\lambda\delta_{ij}}{2\mu + 3\lambda}\tau_{kk}. \qquad \text{(V–3)}$$

Some special cases are:

(*a*) Hydrostatic pressure: If $\tau_{ij} = -p\delta_{ij}$, then the ratio of the pressure p to the fractional decrease in volume $-e_{kk}$ is defined to be the bulk modulus κ. From Eq. (V–2) it follows easily that

$$\kappa = \lambda + \tfrac{2}{3}\mu.$$

(*b*) Pure tension: If $\tau_{11} = T$, all other $\tau_{ij} = 0$, then the ratio τ_{11}/e_{11} is defined to be the elastic modulus (or Young's modulus) E. From Eq. (V–3),

$$E = \frac{\mu(2\mu + 3\lambda)}{\mu + \lambda}.$$

The ratio of the fractional lateral contraction to the linear strain in this case is defined to be Poisson's ratio σ;

$$\sigma = -\frac{e_{22}}{e_{11}} = -\frac{e_{33}}{e_{11}} = \frac{\lambda}{2(\lambda + \mu)}.$$

(*c*) Pure shear: If $\tau_{12} = \tau_{21}$ alone is not zero, then the ratio $\tau_{12}/2e_{12}$ (that is, of shear stress to angle change) is defined to be the shear modulus (or rigidity modulus) G. From Eq. (V–3),

$$G = \mu.$$

By use of these results, any one of the elastic constants E, σ, κ, λ, μ may be expressed in terms of any other two of the constants (Table 5).

TABLE 5. Relations among elastic constants.

	E	σ	κ	μ	λ
E, σ	E	σ	$\frac{E}{3(1-2\sigma)}$	$\frac{E}{2(1+\sigma)}$	$\frac{E\sigma}{(1+\sigma)(1-2\sigma)}$
E, κ	E	$\frac{3\kappa-E}{6\kappa}$	κ	$\frac{3\kappa E}{9\kappa-E}$	$\frac{3\kappa(3\kappa-E)}{9\kappa-E}$
E, μ	E	$\frac{E-2\mu}{2\mu}$	$\frac{\mu E}{3(3\mu-E)}$	μ	$\frac{\mu(E-2\mu)}{3\mu-E}$
E, λ*	E	$\frac{2\lambda}{E+\lambda+\surd}$	$\frac{E+3\lambda+\surd}{6}$	$\frac{E-3\lambda+\surd}{4}$	λ
σ, κ	$3\kappa(1-2\sigma)$	σ	κ	$\frac{3\kappa(1-2\sigma)}{2(1+\sigma)}$	$\frac{3\kappa\sigma}{1+\sigma}$
σ, μ	$2\mu(1+\sigma)$	σ	$\frac{2\mu(1+\sigma)}{3(1-2\sigma)}$	μ	$\frac{2\mu\sigma}{1-2\sigma}$
σ, λ	$\frac{\lambda(1+\sigma)(1-2\sigma)}{\sigma}$	σ	$\frac{\lambda(1+\sigma)}{3\sigma}$	$\frac{\lambda(1-2\sigma)}{2\sigma}$	λ
κ, μ	$\frac{9\kappa\mu}{6\kappa+\mu}$	$\frac{3\kappa-2\mu}{6\kappa+2\mu}$	κ	μ	$\kappa-\frac{2}{3}\mu$
κ, λ	$\frac{9\kappa(\kappa-\lambda)}{3\kappa-\lambda}$	$\frac{\lambda}{3\kappa-\lambda}$	κ	$\frac{3}{2}(\kappa-\lambda)$	λ
μ, λ	$\frac{\mu(3\lambda+2\mu)}{\lambda+\mu}$	$\frac{\lambda}{2\lambda+2\mu}$	$\frac{3\lambda+2\mu}{3}$	μ	λ

* $\surd$ stands for $\surd(E^2 + 9\lambda^2 + 2E\lambda)$.

The fact that a body must allow applied loading to do work on it requires (considering the foregoing special cases) that E, μ, κ be positive; hence

$$-1 < \sigma \leq \tfrac{1}{2},$$
$$\lambda \geq 0.$$

A value of $\frac{1}{2}$ for σ corresponds to incompressibility. An example of a negative σ is cited by Love.[1]

An important consequence of Eqs. (V–2) or (V–3) is that principal axes of stress and strain coincide. This fact suggests an alternative derivation of Eq. (V–3). If it is assumed to begin with that for an isotropic body the principal axes of stress and strain coincide (and this is physically plausible), and if it is assumed that the stress-strain relation is linear so that superposition holds, then the definitions of E and σ imply, for example, that

$$e_{11} = \frac{\tau_{11}}{E} - \sigma\left(\frac{\tau_{22}}{E} + \frac{\tau_{33}}{E}\right),$$

and similarly for e_{22} and e_{33}. This result is identical with Eq. (V–3).

V–2. Summary of Equations

The stress-strain relations (V–2) may be used to express the equilibrium equations in terms of displacements (Navier's equations), or the compatibility equations in terms of stresses (Beltrami-Michell equations). In deducing these latter equations, direct substitution of Eq. (V–3) into Eqs. (IV–35) gives

$$\tau_{ik,jj} - \tau_{ij,jk} - \tau_{kj,ji} + \tau_{jj,ik} - \frac{\sigma}{1+\sigma}(\delta_{ik}\tau_{rr,jj} + \tau_{jj,ik}) = 0, \quad \text{(V–4)}$$

from which the useful result (V–15) may be obtained. Use of Eq. (V–15) in Eq. (V–4) and also of the equilibrium equations then gives Eq. (V–14) below. The final results are:

$$\tau_{ij} = \tau_{ji}, \quad \text{(V–5)}$$

$$T_i = \tau_{ij}n_j, \quad \text{(V–6)}$$

$$\tau_{ij,j} + \rho F_i = \rho a_i, \quad \text{(V–7)}$$

$$e_{ij} = e_{ji} = \tfrac{1}{2}(u_{i,j} + u_{j,i}), \quad \text{(V–8)}$$

$$\omega_{ij} = -\,\omega_{ji} = \tfrac{1}{2}(u_{i,j} - u_{j,i}), \quad \text{(V–9)}$$

$$e_{psk}e_{rij}e_{ik,js} = 0, \quad \text{(V–10)}$$

$$\tau_{ij} = \frac{E}{1+\sigma}\left(e_{ij} + \frac{\sigma}{1-2\sigma}\,e_{kk}\delta_{ij}\right), \quad \text{(V–11)}$$

[1] A. E. H. Love, *A treatise on the mathematical theory of elasticity* (4th ed.; Dover, New York, 1944), p. 163.

$$Ee_{ij} = (1+\sigma)\tau_{ij} - \sigma\delta_{ij}\tau_{kk}, \tag{V-12}$$

$$u_{i,jj} + \frac{1}{1-2\sigma} u_{j,ji} + \frac{1}{G}(\rho F_i - \rho a_i) = 0, \tag{V-13}$$

$$\tau_{ik,jj} + \frac{1}{1+\sigma}\tau_{jj,ik} + (\rho F_i - \rho a_i)_{,k} + (\rho F_k - \rho a_k)_{,i} + \frac{\sigma\delta_{ik}}{1-\sigma}(\rho F_s - \rho a_s)_{,s} = 0, \tag{V-14}$$

$$(1-\sigma)\tau_{ii,jj} - (1+\sigma)\tau_{ij,ji} = 0. \tag{V-15}$$

The acceleration term a_i would be $\partial^2 u_i/\partial t^2$ if Lagrangian variables were being used, and $\partial^2 u_i/\partial t^2$ + (second-order terms in u_i) if Eulerian variables were being used. In linear elasticity, no distinction is made between the two, and so the second-order terms are omitted. Consequently, the equations above are linear (allowing superposition of solutions) and moreover have constant coefficients. Simple solutions can be found (for example, τ_{11} = constant, all other $\tau_{ij} = 0$; appropriate boundary conditions) so that the equation set is not inconsistent. If, furthermore, the equations are accepted as adequately representing the behavior of an elastic body, then physical considerations guarantee the existence of a (unique) solution for any properly posed problem. For certain types of problem, the existence of a sufficiently well-behaved solution has been demonstrated mathematically.[2]

An alternative form of Eq. (V–13) is often useful. Taking the curl of this equation gives

$$\omega_{ij,ss} + \frac{1}{2G}[(\rho F_i - \rho a_i)_{,j} - (\rho F_j - \rho a_j)_{,i}] = 0, \tag{V-16}$$

and alternatively taking the divergence gives

$$u_{i,ijj} + \frac{1-2\sigma}{2-2\sigma}\cdot\frac{1}{G}(\rho F_i - \rho a_i)_{,i} = 0. \tag{V-17}$$

One reason for the usefulness of Eqs. (V–16) and (V–17) is that they do not contain the expression $u_{j,j}/(1-2\sigma)$, which becomes indeterminate as $\sigma \to \frac{1}{2}$; cf. Eq. (V–13). For the case $\sigma = \frac{1}{2}$, Eq. (V–17) may be replaced by $u_{i,i} = 0$. It may be noted that Eqs. (V–16) are not completely independent, because of the fact that

$$\omega_{12,3} + \omega_{23,1} + \omega_{31,2} = 0. \tag{V-18}$$

[2] Some references are given by I. S. Sokolnikoff, *Mathematical theory of elasticity* (2nd ed.; McGraw-Hill, New York, 1956), p. 358.

For the present, only time-independent problems will be considered. Then one type of problem in elasticity is that in which the three u_i are specified on the boundary of a body and it is required to determine the u_i throughout the interior of the body. Here Eqs. (V–13) are appropriate; they form a set of three partial differential equations each of which contains derivatives of all the u_i. The problem is very similar to the familiar Dirichlet problem in which it is required to find a single function U harmonic in a region and possessing prescribed boundary values; it will in fact be subsequently shown that the solution of the elasticity problem can be expressed in terms of harmonic functions. One way of pointing up the analogy in the case of zero body force is to note that Eq. (V–17) then implies $u_{i,ijj} = 0$, so that Eq. (V–13) can be written

$$\left[u_i + \frac{1}{2(1-2\sigma)} x_i u_{s,s}\right]_{,jj} = 0. \qquad \text{(V–19)}$$

The analogy may be pursued further. Suppose that the surface stresses T_i rather than the surface displacements u_i are prescribed. Then

$$\begin{aligned} T_i &= \tau_{ij} n_j \\ &= \frac{E}{1+\sigma}\left[\frac{1}{2}(u_{i,j} + u_{j,i}) + \frac{\sigma}{1-2\sigma}\delta_{ij} u_{k,k}\right] n_j \\ &= \frac{E}{1+\sigma}\left[\frac{\partial u_i}{\partial n} + \omega_{ji} n_j + \frac{\sigma}{1-2\sigma} u_{k,k} n_i\right], \end{aligned} \qquad \text{(V–20)}$$

so that (assuming for the moment that ω_{ij} and $u_{k,k}$ can be determined by other means from the boundary conditions — as indeed they can) a specification of T_i is equivalent to a specification of $\partial u_i/\partial n$. Thus this natural second type of elasticity problem is very similar to the second fundamental (Neumann) problem of potential theory in which the normal derivative of a harmonic function U is prescribed on the boundary. The well-known condition that the Neumann problem be solvable is that the boundary surface integral of $\partial u/\partial n$ should vanish; again there is an analogy to the elasticity problem in that if the surface stresses are everywhere prescribed they must satisfy over-all equilibrium if a time-independent solution is to exist.

It is of course not necessary to attack by way of Eq. (V–13) a problem in which the surface stresses are prescribed; it is often easier to use the lengthier set of Eqs. (V–7) and (V–14). This is a set of nine equations for the determination of the six τ_{ij}; however, as pointed out in Sec. IV–5, only three of the six Eqs. (V–14) are strongly independent, the other three giving only some information regarding arbitrary constants or functions of integration. It may incidentally be noted that any stress-state for which each τ_{ij} is a linear function of the coordinates automatically satisfies Eq. (V–14) for the case of zero or constant body force.

A more general "boundary-value" problem is one in which u_i is specified over part of the boundary and T_i over the remainder. It is also possible to specify one component of displacement (say normal) over part of the surface, and a conjugate component of stress (here tangential) over the same part of the surface; still more generally, a linear combination of stress and displacement (as in spring loading) may be specified over part of the surface.

As a general rule, a problem for which the boundary conditions can be visualized physically will not lead to conceptual difficulties during the process of obtaining a solution. It is, however, often convenient to idealize some aspects of a problem, even to such an extent that difficulties of existence or uniqueness may occur. For example, a body may be considered to be infinite or semi-infinite in extent, whereas any real body is finite; a problem may be assumed to be essentially two-dimensional, whereas any real problem is three-dimensional; a load may be considered to be concentrated at a point, whereas any real load is applied over a nonzero area. A reasonably safe procedure in such cases is to consider a sequence of real problems in which the idealized problem is approached as a limiting case.

There are few fields in which the interplay of physical reasoning and mathematical analysis is as important as in elasticity. Usually, the problems met with are too complicated to solve directly, and so recourse must be had to approximate or numerical methods. For example, there are certain variational principles (Chapter VII) which are equivalent to the equations of elasticity, and these principles may be used to determine the best form of a guessed solution; the procedure is usually efficient only if the guessed solution is not too far from the true one. Also, it may happen that it is possible to solve exactly a problem which in some sense is not too far from the actual one. For example, the elasticity problem for the torsion of a long cylinder of arbitrary cross section and with shearing stresses distributed over the ends in a certain manner has been solved; if the actual end stress distribution differs in detail from this but produces the same resultant twisting moment, it is reasonable to expect the solutions of the two problems to differ appreciably only near the ends — it being argued that, as far as material well removed from the ends is concerned, only the statical resultant of the applied forces should be important. This *St. Venant Principle*, whereby it is permissible in certain cases to replace loads by statical equivalents, has been investigated mathematically for some simple situations; however, the best criterion of validity in a practical problem is again to be found in physical intuition.

If the body force has a potential so that $\rho F_i = \phi_{,i}$ (as is the case for a constant-density body in a gravitational field or in uniform rotation, F_i being the inertial equivalent of central acceleration in this latter example), then Eq. (V–16) implies that

$$\omega_{ij,ss} = 0,$$

so that each component of rotation is harmonic. If ϕ is harmonic, then so are $u_{i,i}$ and τ_{ii}; further, the u_i and τ_{ij} are in this case individually biharmonic ($u_{i,jjss} = 0$). These results hold even if the body is incompressible. Incompressibility sometimes leads to a difficulty in that τ_{ij} cannot be determined from given e_{ij} by Eq. (V–11), because the term $e_{kk}/(1 - 2\sigma)$ is indeterminate; no such difficulty occurs in the converse equation, Eq. (V–12). Denoting the indeterminate part of Eq. (V–11) by $p\delta_{ij}$, so that

$$\tau_{ij} = \frac{E}{1+\sigma} e_{ij} + p\delta_{ij},$$

it follows that

$$\tau_{ij,j} = \frac{E}{1+\sigma} e_{ij,j} + p_{,i},$$

from which

$$p_{,i} = -\rho F_i - \frac{E}{1+\sigma} e_{ij,j}. \tag{V–21}$$

Consequently a knowledge of body force and of e_{ij} is enough to determine p within an arbitrary constant. If $e_{ij,j}$ is replaced by $\frac{1}{2}u_{i,jj}$ (since $u_{j,j} = 0$ here), the converse interpretation of Eq. (V–21) for the case of zero body force is that the Laplacian of the displacement vector is the gradient of some function (the same result follows from Eq. (V–16), which for $\rho F_i = 0$ implies that the curl of the Laplacian of the displacement vector vanishes).

Before proceeding with the derivation of some general theorems, it is of interest to consider two analogies from other fields. Equation (V–13) is

$$u_{i,jj} + \frac{1}{1-2\sigma} u_{j,ji} + \frac{1}{G} \rho F_i = \frac{1}{G} \rho \frac{\partial^2 u_i}{\partial t^2},$$

which may be compared with the equation governing the velocity v_i in a slowly moving Newtonian fluid of viscosity μ:

$$v_{i,jj} + \frac{1}{3} v_{j,ji} + \frac{1}{\mu}(\rho F_i - p_{,i}) = \frac{1}{\mu}\rho\frac{\partial v_i}{\partial t}.$$

Second, a powerful general method of obtaining analogies is to replace given differential equations with equivalent difference equations (that is, a differential quotient dy/dx is replaced by the ratio $(y_2 - y_1)/\delta x$, where y_1, y_2 are the values of y at adjoining points separated by a distance δx); an electric circuit may then always be devised so that the currents or voltages correspond to the values of the desired quantities at the mesh points. In the case of elasticity, many such circuits are possible; a particularly simple one has been given by Kron.[3]

V–3. Removal of Body Force. Singularities

Consider an incompressible body at rest and with $F_i = 0$. Then Eq. (V–16) implies that the curl of $u_{i,ss}$ vanishes, so that by Sec. I–13 $u_{i,ss} = \phi_{,i}$ for some ϕ. Kelvin noted in 1847 in connection with some work on magnetic effects that a special solution of the equations

$$\operatorname{curl}(u_{i,ss}) = 0, \quad u_{i,i} = 0$$

was given by

$$u_i = \frac{1}{2}\left(\frac{p_j x_j}{r}\right)_{,i} - \frac{p_i}{r},$$

where p_i is an arbitrary constant vector and where the distance r from the origin is given by $r = (x_i x_i)^{\frac{1}{2}}$. This form of u_i satisfies the two equations everywhere except at the origin, where there is a singularity. Note that the quantity $(1/r)$ is harmonic except at the origin, and that the Laplacian of the product of two functions is given by

$$(\alpha\beta)_{,ss} = \alpha_{,ss}\beta + \alpha\beta_{,ss} + 2\alpha_{,s}\beta_{,s}.$$

Subsequently, Kelvin discovered that the slight modification

$$u_i = \frac{c_1}{2}\left(\frac{p_j x_j}{r}\right)_{,i} - c_2\frac{p_i}{r}$$

satisfies Navier's equations (V–13), provided that $c_2 = c_1(2 - 2\sigma)$, so that the restriction of incompressibility is unnecessary. Investigating the

[3] See, for example, W. W. Soroka, *Analog circuits for computation and simulation* (McGraw-Hill, New York, 1954), p. 322.

nature of the singularity, he found that if a spherical region surrounding the origin were removed then the stresses required on the interior spherical surface by the foregoing displacements were statically equivalent to a single force in the direction of p_i and of magnitude independent of the size of the sphere (in checking this calculation, note that the inward unit normal is given by $n_i = -x_i/r$). Allowing the sphere to shrink to zero then results in a concentrated force at the origin; if the constants are adjusted so as to correspond to a concentrated force of unit magnitude in the positive x_j-direction, the result (which holds also for the special case $\sigma = \frac{1}{2}$) is

$$u_i^{(j)} = \frac{1}{4\pi E}\left(\frac{1+\sigma}{1-\sigma}\right)\left[(2-2\sigma)\frac{\delta_{ij}}{r} - \frac{1}{2}r_{,ij}\right]. \tag{V–22}$$

The order of the displacement singularity is $1/r$; the corresponding stresses have singularities of order $1/r^2$. The dilation is given by

$$u_{i,i}^{(j)} = \frac{1}{4\pi E}\left(\frac{1+\sigma}{1-\sigma}\right)(1-2\sigma)\left(\frac{1}{r}\right)_{,j}. \tag{V–23}$$

If now there is a distribution of force throughout the body, a displacement vector satisfying Eqs. (V–13) (that is, the equilibrium equations in terms of displacements) may be obtained by considering the body force as a collection of concentrated forces to each of which Eq. (V–22) is applicable; in the limit, the summation becomes an integral and we have

$$u_i = \frac{1}{4\pi E}\left(\frac{1+\sigma}{1-\sigma}\right)\left[(2-2\sigma)\int\frac{\rho F_i(\xi)}{r(x;\xi)}\,dV - \frac{1}{2}\int\frac{\partial}{\partial x_i}\left(\frac{x_j-\xi_j}{r(x;\xi)}\right)\rho F_j(\xi)\,dV\right], \tag{V–24}$$

where the integration is carried out over (ξ_1, ξ_2, ξ_3) and where

$$r(x;\xi) = [(x_1-\xi_1)^2 + (x_2-\xi_2)^2 + (x_3-\xi_3)^2]^{\frac{1}{2}}.$$

Since integration is a limiting process in which each element of volume — and so also the total body force on that element of volume — shrinks to zero, it may be expected that (for smoothly varying F_i) the resulting u_i has no singularities. If now the body force is prescribed in some elasticity problem, Eq. (V–24) may be used to generate a displacement vector which is in equilibrium with this body force; the difference between the required final solution and this displacement vector then must correspond to zero body force (the boundary conditions satisfied by the difference vector are of course not the same as the original boundary conditions) and consequently Eq. (V–24) provides a method for removing body force from consideration.

The method may be put in an alternative form. Define

$$Q_i = \frac{(1+\sigma)}{2\pi E} \int \frac{\rho F_i(\xi)}{r(x;\xi)}\, dV,$$

so that by a standard formula in potential theory [4]

$$Q_{i,ss} = -\frac{2(1+\sigma)}{E}\, \rho F_i.$$

Next define

$$M = -\frac{(1+\sigma)}{8E\pi(1-\sigma)} \int \frac{x_j - \xi_j}{r(x;\xi)}\, \rho F_j(\xi)\, dV,$$

so that [4]

$$M_{,ss} = -\frac{(1+\sigma)}{8E\pi(1-\sigma)} \left\{ (-4\pi)[(x_j-\xi_j)\rho F_j(\xi)] \Big|_{\xi=x} + 2\int \delta_{js} \left(\frac{1}{r}\right)_{,s} \rho F_j\, dV \right\}$$

$$= 0 - \frac{1}{2-2\sigma}\, Q_{s,s}.$$

Consequently the displacement vector

$$u_i = Q_i + M_{,i}, \qquad \text{(V–25)}$$

where Q_i and M satisfy the simple Poisson equations

$$Q_{i,ss} = -\frac{1}{G}\, \rho F_i,$$

$$M_{,ss} = -\frac{1}{2-2\sigma}\, Q_{s,s},$$

is a particular solution of Eq. (V–13) for the body force F_i (and so may be used to remove that body force). This statement may be verified by direct substitution of Eq. (V–25) into Eq. (V–13). Note that boundary conditions in the solutions of the two Poisson equations are unimportant; *any* solutions Q_i, M are permissible. If stresses rather than displacements are of primary interest (that is, if Eqs. (V–7) and (V–14) are being used), then the stresses corresponding to Eq. (V–25) may again be used to remove body forces from consideration. The essential result of this section is that in subsequent discussions of the structure of Eqs. (V–5)

[4] O. D. Kellogg, *Foundations of Potential Theory*, Ungar, 1929, pp. 152, 153. Or see Sec. VI–1.

to (V–15) for a body at rest, it is permissible without loss of generality to set $F_i = 0$.

Returning now to the concentrated force singularity (V–22), there are a number of other types of singularity that can be constructed from it.

(1) *Double force without moment.* At $(x_1 = h/2, x_2 = x_3 = 0)$, place a force of magnitude P/h and directed along the positive x_1-axis; at $(x_1 = -h/2, x_2 = x_3 = 0)$ place a force P/h directed along the negative x_1-axis. Then

$$u_i = \frac{P}{h}\left[u_i^{(1)}\left(x_1 - \frac{h}{2}, x_2, x_3\right) - u_i^{(1)}\left(x_1 + \frac{h}{2}, x_2, x_3\right)\right],$$

and as $h \to 0$ the limiting value of the resulting displacement is

$$u_i = -P\frac{\partial}{\partial x_1}u_i^{(1)}. \tag{V–26}$$

(2) *Center of expansive pressure.* Repeat the process of (1) above for the other two axes, and add the three results together to give

$$\begin{aligned} u_i &= -P\left[\frac{\partial u_i^{(1)}}{\partial x_1} + \frac{\partial u_i^{(2)}}{\partial x_2} + \frac{\partial u_i^{(3)}}{\partial x_3}\right] \\ &= -\frac{P}{4\pi E}\left(\frac{1+\sigma}{1-\sigma}\right)(1-2\sigma)\left(\frac{1}{r}\right)_{,i}. \end{aligned}$$

For a small sphere surrounding the origin, the stresses are found to correspond to a uniform outward pressure of intensity

$$p = \frac{2P(1-2\sigma)}{1-\sigma}\frac{1}{4\pi R^3},$$

where R is the radius of the sphere. Note that $\frac{4}{3}\pi R^3 p$ is independent of R. If $\frac{4}{3}\pi R^3 p$ is denoted by K, the displacement is

$$u_i = -\frac{3(1+\sigma)}{8\pi E}K\left(\frac{1}{r}\right)_{,i}. \tag{V–27}$$

It follows at once that $u_{i,i} = 0$ and that $\omega_{ij} = 0$.

(3) *Concentrated moment.* At $(x_1 = h/2, x_2 = x_3 = 0)$ place a force M/h in the direction of the positive x_2-axis; at $(x_1 = -h/2, x_2 = x_3 = 0)$ place an equal but oppositely directed force. Then

$$u_i = \frac{M}{h}\left[u_i^{(2)}\left(x_1 - \frac{h}{2}, x_2, x_3\right) - u_i^{(2)}\left(x_1 + \frac{h}{2}, x_2, x_3\right)\right],$$

and as $h \to 0$ the limiting value of the displacement corresponding to a concentrated couple M about the x_3-axis becomes

$$u_i = -M\frac{\partial u_i^{(2)}}{\partial x_1}.$$

This expression is not particularly symmetrical. If, however, similar forces parallel to the x_1-axis are allowed to act at $(x_2 = h/2, x_1 = x_3 = 0)$, $(x_2 = -h/2, x_1 = x_3 = 0)$, and the result is added to the previous one so as to give a concentrated couple of amount $2M$ about the x_3-axis, the displacement is

$$u_i = M\left(\frac{1+\sigma}{2\pi E}\right) e_{3it}\left(\frac{1}{r}\right)_{,t}. \tag{V-28}$$

Again, $u_{i,i} = 0$.

(4) *Semi-infinite line of centers of expansion.* Let a uniform distribution of centers of expansion, of type (2) above, be placed on the line $0 \leq x_3 \leq \infty$. By integration, the displacements are found to be

$$u_i = C[\ln(r - x_3)]_{,i},$$

where C is proportional to the intensity of distribution. Boussinesq described this displacement vector as a singular solution of "the second type," and used it in conjunction with Eq. (V–22) to give the solution for a concentrated surface force acting normal to the plane bounding a semi-infinite body.

(5) *Concentrated force normal to a plane boundary.* Consider an elastic medium occupying the half-space $x_3 \leq 0$; at the origin of coordinates, apply a concentrated force in the positive x_3-direction. If the singularity (V–22) with $j = 3$ is first tried alone, it will be found that the stresses over a hemispherical surface surrounding the origin have a resultant only in the x_3-direction; however, the plane surface is not free from stress but requires

$$\tau_{31} = \frac{AE}{2(1+\sigma)}(1-2\sigma)\left(-\frac{x_1}{r^3}\right),$$

$$\tau_{32} = \frac{AE}{2(1+\sigma)}(1-2\sigma)\left(-\frac{x_2}{r^3}\right),$$

$$\tau_{33} = 0,$$

where $A = \dfrac{1}{4\pi E}\left(\dfrac{1+\sigma}{1-\sigma}\right)$.

Now the stress on $x_3 = 0$ corresponding to (4) above is

$$\tau_{31} = \frac{CE}{(1+\sigma)}\left(\frac{x_1}{r^3}\right),$$

$$\tau_{32} = \frac{CE}{(1+\sigma)}\left(\frac{x_2}{r^3}\right),$$

$$\tau_{33} = 0.$$

Thus a choice of ratio of C to A given by

$$C - \tfrac{1}{2}A(1-2\sigma) = 0$$

results in zero surface stress; the two displacements in question require concentrated loads in the positive x_3-direction of magnitudes $\frac{1}{2}$ and $-2\pi EC/(1+\sigma)$ respectively, so that the net force would be

$$\frac{1}{2} - \frac{1-2\sigma}{4(1-\sigma)} = \frac{1}{4(1-\sigma)}.$$

Consequently a unit concentrated load in the positive x_3-direction, applied at the origin to the half-space $x_3 \leq 0$, will correspond to a displacement of

$$u_i = \frac{1}{4\pi G}\left[\frac{x_3 x_i}{r^3} + (3-4\sigma)\frac{\delta_{i3}}{r} + (1-2\sigma)\frac{x_i - \delta_{3i} r}{r(r-x_3)}\right]. \qquad \text{(V-29)}$$

Again, the displacement has a $1/r$ singularity, and the stress a $1/r^2$ singularity. Also,

$$u_{i,i} = \frac{1}{2\pi G}(1-2\sigma)\left(\frac{1}{r}\right)_{,3}.$$

If a concentrated normal force is applied to a locally spherical portion of the surface of a body, then it is found that Eq. (V–29) does not correctly describe the character of the displacement singularity, but that a lower-order singularity must be added.[5]

(6) *Concentrated force parallel to a plane boundary.* In a somewhat similar manner, the displacement corresponding to a unit force in the positive x_1-direction applied at the origin to the half-space $x_3 \leq 0$ is found to be

$$u_i = \frac{1}{4\pi G r}\left\{\delta_{1i} + \frac{x_1 x_i}{r^2} + \frac{1-2\sigma}{(r-x_3)^2}\left[r(r-x_3)\delta_{1i} - x_1 x_i + x_1(2x_3 - r)\delta_{3i}\right]\right\}. \qquad \text{(V-30)}$$

Note that

$$u_{i,i} = \frac{(1-2\sigma)}{2\pi G}\left(\frac{1}{r}\right)_{,1}.$$

It should be noted that entirely different displacement (and stress) singularities may correspond to the same resultant force or moment. For example, a superposition of the two u_i given by Eqs. (V–22) and (V–27) yields a displacement which corresponds to a concentrated unit body force, yet which differs (by a singularity of higher order) from that of Eq. (V–27). Similarly, another displacement vector which equilibrates a concentrated normal force applied to the planar surface of a half-space may be obtained by superposing the solutions, as given by Eq. (V–29), of three concentrated normal forces, namely, $1/h^2$ at $(h, 0, 0)$ and at

[5] E. Sternberg and R. A. Eubanks, "On the singularity at a concentrated load applied to a curved surface," Technical Report to ONR (Department of Mechanics, Illinois Institute of Technology, 1953).

$(-h, 0, 0)$, and $-2/h^2$ at $(0, 0, 0)$. As $h \to 0$, these forces have no overall resultant, so that the resulting displacement, which is the second partial derivative with respect to x_1 of u_i in Eq. (V–29), may be added to u_i of Eq. (V–29) without altering the character of the required concentrated load. Again, the new u_i has a stronger singularity than the preceding one.

In either case, the choice of the correct singularity must be made on physical grounds. If a sphere (or, for a boundary point, a partial sphere) is drawn around the point at which the concentrated load is applied, then the requisite stress on this spherical surface will be distributed differently for each type of singularity, and so the way in which the physical load is actually applied will govern the character of the singularity. Whenever one tries to apply a concentrated load to a body, the stresses in the neighborhood of the load exceed the elastic limit, so that the material there flows plastically and tends to distribute the load fairly uniformly around the point of load application; this means that the types of singularity observed experimentally are those of lowest order. For example, the triplet of surface forces considered above would require an infinite oscillation of forces, but any such effect would be smoothed out in the plastic zone so that such a singularity would not be observed physically.

In discussing Eq. (V–24), it was remarked that integration tended to cancel singularities. This is in general true if the applied loads are smoothly (continuously) distributed, but not if there are discontinuities. For example, a uniform tangential shear stress of 1 lb/in.^{2} in the x_1-direction, applied to a square finite region of the bounding plane surface of the half-space $x_3 \geq 0$, will result in a displacement vector which has singularities at the borders of this square region.

Singularities cannot exist physically; they are wiped out by plastic flow processes, as mentioned above. They are primarily of interest as a mathematical tool and there is no mathematical objection to having a function (u_i) attain an infinite value at a certain point.

V–4. Uniqueness

Consider the strictly mathematical problem represented by Eqs. (V–5) to (V–15), with $a_i = 0$. For a given region and for prescribed (reasonable) boundary conditions, it is easy to prove that in general the solution is unique. Assume first that the region is finite and that there are no singularities.

If then two solutions, denoted by superscripts (1) and (2), for a given problem are possible, define

$$u_i = u_i^{(1)} - u_i^{(2)},$$
$$e_{ij} = \tfrac{1}{2}(u_{i,j} + u_{j,i}) = e_{ij}^{(1)} - e_{ij}^{(2)},$$
$$\tau_{ij} = \tau_{ij}^{(1)} - \tau_{ij}^{(2)},$$

and note that

$$\tau_{ij,j} = \tau_{ij,j}^{(1)} - \tau_{ij,j}^{(2)} = -\rho F_i + \rho F_i = 0.$$

The quantity

$$\int e_{ij}\tau_{ij}\,dV$$

is nonnegative, as can be seen by expressing τ_{ij} in terms of e_{ij} by means of Eq. (V–11) (in the special case $\sigma = \frac{1}{2}$, write $\tau_{ij} = Ee_{ij}/(1+\sigma) + \delta_{ij}p$ and note that $e_{ii} = 0$); it vanishes only if $e_{ij} = 0$. Expanding,

$$\begin{aligned}\int e_{ij}\tau_{ij}\,dV &= \int \tfrac{1}{2}(u_{i,j} + u_{j,i})\tau_{ij}\,dV \\ &= \int u_{i,j}\tau_{ij}\,dV \\ &= \int (u_i\tau_{ij})_{,j}\,dV - \int u_i\tau_{ij,j}\,dV \\ &= \int u_i\tau_{ij}n_j\,dS \\ &= \int (u_i^{(1)} - u_i^{(2)})(T_i^{(1)} - T_i^{(2)})\,dS; \qquad \text{(V–31)}\end{aligned}$$

and, if the prescribed boundary conditions are such that this last integral vanishes, then this will imply that $e_{ij} = 0$ and so that the only difference between solutions (1) and (2) is a rigid-body motion (and if the displacement is known on part of the boundary, there cannot even be a difference in rigid-body motion between the solutions). Typical conditions under which the integrand vanishes are:

(1) u_i prescribed over the entire surface (so that $u_i^{(1)} = u_i^{(2)}$ on S),

(2) T_i prescribed over the entire surface (but satisfying over-all equilibrium with F_i in order that the body remain at rest),

(3) u_i prescribed on part of the surface, T_i on the remainder,

(4) normal component of displacement and tangential component of stress vector (or vice versa) prescribed over the whole surface, or over part of the surface with one of the previous conditions over the rest of the surface.

Uniqueness may also be proved for problems involving spring loading over part of the boundary. To illustrate the method, let

$$fu_n + gT_n = h \tag{V–32}$$

for part of the boundary, where u_n and T_n denote prescribed values of normal displacement and stress vector, and where f, g, h are given functions which may vary from point to point on the surface. Any real spring loading will require f and g to have the same sign when nonzero. For the same part of the boundary, let the tangential component of stress, T_t, be prescribed to be zero; for the remainder of the boundary, let one of the four types of boundary condition described above be prescribed. Then the difference solution satisfies

$$fu_n + gT_n = 0$$

for the part of the boundary where Eq. (V–32) is applicable, and the resulting contribution of the right-hand side of Eq. (V–31) is clearly negative unless the difference solution vanishes. Consequently, uniqueness is again proved.

It may be noted in the proof of the uniqueness theorem that each of solutions (1) and (2) corresponds to the same body shape, and consequently there is no contradiction to the known fact that a body under prescribed loading may occupy two or more different equilibrium positions (as in stability problems). What the theorem does say is that any solution of the *mathematical* problem represented by Eqs. (V–5) to (V–15) for a *given* region is unique.

If part of the region extends to infinity, then it is usually required that the stresses should approach zero at infinity. However, the rate of approach to zero may not be sufficiently rapid to make the right-hand side of Eq. (V–31) approach zero (the size of the surface of integration usually approaches infinity), so that uniqueness of solution may not hold. Each such problem must be considered on its own merits.

Precisely similar remarks hold if singularities are allowed; a limiting sequence of surfaces enclosing the singularity must be considered. It has been pointed out by Sternberg and Eubanks [6] that it is usually sufficient to specify the order of the singularity in order to ensure uniqueness.

There is another situation in which the uniqueness theorem cannot be applied directly. Suppose that in a ring of metal a thin wedge-shaped

[6] E. Sternberg and R. A. Eubanks, "On the concept of concentrated loads," *J. Rational Mech. and Analysis* *4* (1955), 135.

portion bounded by two transverse planes is removed, and the two resulting faces are then forced together and welded. The resulting stresses are not zero even if there are no surface tractions or body forces applied, and this would be a contradiction to the uniqueness theorem if it were applicable. This type of *dislocation*, resulting in initial stress, is best handled by treating the weld as a discontinuity[7] across which certain mutual shear and normal stresses act; the combination of these interface stresses and any external loading may then be thought of as being applied to an initially stress-free body. Although simple and plausible methods (developed by Weingarten and Volterra) may be used to determine the solution in such comparatively easy cases as the present example, the general problem of determining the intersurface stresses requires the solution of a set of integral equations and is mathematically tedious. If the precise method whereby the dislocation was achieved is not known, it may in principle be determined by observing the behavior of the body as various loadings are applied. Initial stresses may also arise from unequal heating of the body, or from forming processes in which a part of the material is prevented from returning to its stress-free state by the resistance of adjoining material. If the heating or forming distortion is known, then the mathematical problem of analyzing the behavior of the body may be simplified by the incorporation of an artificial body force (see Chapter VIII).

An interesting way of using the uniqueness theorem is to show that a given set of elastic equations is complete. Consider for example Eqs. (V–16) and (V–17), for the case $\sigma = \frac{1}{2}$; for simplicity, let $F_i = 0$, and let the body be at rest. These equations were derived from the general equations of elasticity, and so must be satisfied by any solution of these general equations; on the other hand, it is not clear that any solution of these secondary equations will satisfy the general equations. This fact would, however, follow if it were known that the secondary equations had a unique solution. We have

$$u_{i,jss} - u_{j,iss} = 0,$$
$$u_{i,i} = 0.$$

[7] If the discontinuity is thought of as coinciding with the $x_3 = 0$ plane, then of the six stresses only τ_{31}, τ_{32}, τ_{33} need be continuous across the plane. The effect of the welding may be to smooth out discontinuities in the other stress components, but this is accompanied by unknown stress-relieving metallurgical changes in the metal; it is usually preferable to treat the weld as an actual discontinuity and to use St. Venant's principle to justify the accuracy of the solution at points not too close to the weld.

We know that there exists at least one solution corresponding to prescribed surface displacements; if there existed two different solutions, then the difference solution would also satisfy these equations and moreover would correspond to zero surface displacement. For the difference solution,

$$\int u_{i,j}u_{i,j}\,dV = \int (u_i u_{i,j})_{,j}\,dV - \int u_i u_{i,jj}\,dV$$
$$= -\int u_i u_{i,jj}\,dV,$$

since the first integral vanishes after use of the divergence theorem. But the first set of equations implies that $u_{i,jj} = \phi_{,i}$ where ϕ is some function (for the curl of the vector $u_{i,ss}$ vanishes) and so

$$\int u_{i,j}u_{i,j}\,dV = -\int u_i\phi_{,i}\,dV = -\int (u_i\phi)_{,i}\,dV,$$

since $u_{i,i} = 0$. The right-hand side vanishes (after transforming to a surface integral) and so therefore does the left-hand side, which is however the integral of a sum of squares. Consequently $u_{i,j} = 0$, so u_i is at most a constant, which must in fact equal zero. This proves the desired result.

V–5. Integral Results

First, an expression for the stored energy of an elastic body will be deduced. Consider any volume element dV of the body; the local stress and strain tensors have values τ_{ij} and e_{ij}. Let local principal axes be used, so that only the main-diagonal terms of these two tensors do not vanish. One way of bringing the volume element from its initial into its final state is to apply stresses which grow uniformly and at the same relative rate from zero to their final values; the work done in this process (which must equal the work done in the actual process, since stored internal energy is a state function and so depends only on the final state) is clearly

$$\tfrac{1}{2}(\tau_{11}e_{11} + \tau_{22}e_{22} + \tau_{33}e_{33})\,dV.$$

In an arbitrary axis system, this expression is

$$\tfrac{1}{2}\tau_{ij}e_{ij}\,dV,$$

and integrating over the whole body gives the elastic energy as

$$\mathcal{E} = \tfrac{1}{2}\int \tau_{ij}e_{ij}\,dV. \qquad \text{(V–33)}$$

Thermal effects have been neglected here; the more exact calculation of stored energy to be given subsequently shows that Eq. (V–33) is usually sufficiently correct. In any event, this equation may be used as a definition of a quantity $\mathcal{E}$, and various future identities involving $\mathcal{E}$ are true independently of how closely $\mathcal{E}$ agrees with the physical notion of stored energy.

One immediate identity is obtained by expansion of e_{ij}, using symmetry of τ_{ij}:

$$\begin{aligned}\mathcal{E} &= \tfrac{1}{2}\int \tau_{ij}u_{i,j}\,dV \\ &= \tfrac{1}{2}\int (\tau_{ij}u_i)_{,j}\,dV - \tfrac{1}{2}\int \tau_{ij,j}u_i\,dV \\ &= \tfrac{1}{2}\int T_iu_i\,dS + \tfrac{1}{2}\int \rho F_iu_i\,dV,\end{aligned} \tag{V–34}$$

which has an obvious interpretation in terms of equating external work done to stored energy.

The quantity $\mathcal{E}$ may be expressed entirely in terms of e_{ij}, or entirely in terms of τ_{ij}, by use of the stress-strain laws. The resulting quantities — equal numerically, but of different functional form — will be referred to as the strain and stress energies respectively:

$$\begin{aligned}\mathcal{E}_e &= G\int \left[e_{ij}e_{ij} + \frac{\sigma}{1-2\sigma}\,e_{kk}^2\right] dV, \\ \mathcal{E}_\tau &= \frac{1}{2E}\int [(1+\sigma)\tau_{ij}\tau_{ij} - \sigma\tau_{kk}^2]\,dV.\end{aligned} \tag{V–35}$$

Each form is positive definite, vanishing only when the e_{ij} or τ_{ij} tensor vanishes identically.

Consider now a fixed region V (surface S), and let $e_{ij}^{(1)}$ and $\tau_{ij}^{(1)}$ be any strain and stress tensors related by the stress-strain laws. The corresponding body force and boundary conditions are for the moment of no interest. Let $e_{ij}^{(2)}$ and $\tau_{ij}^{(2)}$ be another such pair. Then the useful identity

$$\int \tau_{ij}^{(1)}e_{ij}^{(2)}\,dV = \int \tau_{ij}^{(2)}e_{ij}^{(1)}\,dV \tag{V–36}$$

may be proved at once by use of the stress-strain laws.

If next the corresponding body forces and surface tractions are denoted by $F_i^{(1)}$, $T_i^{(1)}$ and $F_i^{(2)}$, $T_i^{(2)}$ respectively, then expansion of the e_{ij} in each side of Eq. (V–36) (cf. the derivation of Eq. (V–31)) yields the *reciprocal theorem* of Betti and Rayleigh:

$$\int T_i^{(1)}u_i^{(2)}\,dS + \int \rho F_i^{(1)}u_i^{(2)}\,dV = \int T_i^{(2)}u_i^{(1)}\,dS + \int \rho F_i^{(2)}u_i^{(1)}\,dV,$$

that is, the work that would be done by the external forces (1) acting through the displacements (2) is equal to that done by the external forces (2) acting through the displacements (1).

It is often useful to list the deflections at a number of points on an elastic body due to a unit force applied at each of these points; the reciprocal theorem then implies that there is a symmetry in these *influence coefficients*. Another application of the theorem is obtained by choosing the state (2) to be a very simple one, such as that which results from uniform external pressure; the theorem then yields interesting properties of the solution (1) (for instance, in the pressure example, the volume change of the body due to some other loading could easily be calculated). Still another use of the reciprocal theorem is in obtaining general solutions of elasticity problems, as in Chapter VI.

The reciprocal theorem may also be used to prove that the addition of new material to an elastic body makes it more rigid, provided that there are no body forces. Let the surface of a body consist of two parts, S and s. On S apply tractions T_i, resulting in a displacement pattern $u_i^{(1)}$. Now consider a new problem, in which material has been added to s; when the same T_i are applied to S, the new displacement is $u_i^{(2)}$. Let the stresses exerted by the new material on the old across s be t_i. Then the stresses t_i alone if applied on s to the old material would give displacements $u_i^{(2)} - u_i^{(1)}$; by Eq. (V–34),

$$\int t_i(u_i^{(2)} - u_i^{(1)})\,ds > 0.$$

As far as the new material is concerned, the stresses $-t_i$ correspond to displacements $u_i^{(2)}$, so that

$$\int (-t_i)u_i^{(2)}\,ds > 0.$$

Use of the preceding inequality now gives

$$\int t_i u_i^{(1)}\,ds < 0.$$

But by the reciprocal theorem,

$$\int t_i u_i^{(1)}\,ds = \int T_i(u_i^{(2)} - u_i^{(1)})\,dS,$$

so that

$$\int T_i u_i^{(2)}\,dS < \int T_i u_i^{(1)}\,dS,$$

which is the required result. Although the body is made more rigid by the addition of new material, it must not be inferred that the stresses are everywhere decreased; it is, after all, possible to put a surface notch in the new material so as to result in an infinite stress concentration.

V–6. Elasticity Equations in Curvilinear Coordinates

For reference purposes, it is convenient to list the various equations of elasticity in a form appropriate to more general coordinates than Cartesian. Let (y_1, y_2, y_3) be a curvilinear coordinate system, where each y_i is some function of the Cartesian coordinates (x_1, x_2, x_3), and vice versa. A line along which only one y_i changes is called a parametric curve. Through each point in space pass three such parametric curves; if the curves always intersect orthogonally, the coordinate set is said to be orthogonal. Nonorthogonal sets have very occasionally been used in elasticity (for example, in "follow-the-fiber" methods in elastic stability), but it appears that there are in general no resultant advantages; consequently we will restrict ourselves to the much simpler case of orthogonal coordinates.

Throughout this section — and in this section only — the summation convention of Chapter I will *not* be used.

The easiest way in which to characterize the (y_i) set is to express the square of the distance between adjoining points by

$$ds^2 = h_1^2\,dy_1^2 + h_2^2\,dy_2^2 + h_3^2\,dy_3^2 \qquad \text{(V–37)}$$

where the h_i are certain functions of the y_i. For example, if cylindrical coordinates (r, θ, z) were being used, then

$$\begin{aligned} y_1 &= r, & y_2 &= \theta, & y_3 &= z, \\ h_1 &= 1, & h_2 &= r, & h_3 &= 1. \end{aligned}$$

Note that as a consequence of orthogonality there are no mixed terms in Eq. (V–37) of the typical form $h_{12}\,dy_1\,dy_2$.

Consider first the various formulas of vector analysis of Chapter I. Let **A** be a vector which may vary in magnitude and direction from point to point. At some point P, the projections of the vector onto straight lines tangent to the parametric curves at P will be denoted by A_1, A_2, A_3 and will be called the curvilinear components of **A**. The magnitude of **A** is still given by

$$|\mathbf{A}|^2 = A_1^2 + A_2^2 + A_3^2;$$

however, where the dot or cross product of two vectors acting at points P, Q respectively are required, account must be taken (in an obvious manner) of the fact that the local components are referred to differently oriented systems. If P and Q coincide, the usual formulas hold.

Let ϕ be any scalar. The gradient of ϕ is defined precisely as before with respect to an over-all Cartesian system; the resulting vector has at any point components along the parametric curves which are easily found to be

$$\nabla\phi = \left(\frac{1}{h_1}\frac{\partial\phi}{\partial y_1}, \frac{1}{h_2}\frac{\partial\phi}{\partial y_2}, \frac{1}{h_3}\frac{\partial\phi}{\partial y_3}\right), \tag{V–38}$$

a result which is very reasonable because of the fact that $h_1\, dy_1$, and so forth, are physical distance elements in the parametric directions. If the divergence of a vector is transformed to curvilinear coordinates (either by direct substitution or by use of Gauss's theorem to define the divergence), the result is

$$\nabla\cdot\mathbf{A} = \frac{1}{h_1h_2h_3}\left[\frac{\partial}{\partial y_1}(A_1h_2h_3) + \frac{\partial}{\partial y_2}(A_2h_1h_3) + \frac{\partial}{\partial y_3}(A_3h_1h_2)\right]. \tag{V–39}$$

Similarly the curl becomes

$$\nabla\times\mathbf{A} = \left\{\frac{1}{h_2h_3}\left[\frac{\partial}{\partial y_2}(h_3A_3) - \frac{\partial}{\partial y_3}(h_2A_2)\right], \frac{1}{h_1h_3}\left[\frac{\partial}{\partial y_3}(h_1A_1) - \frac{\partial}{\partial y_1}(h_3A_3)\right], \frac{1}{h_1h_2}\left[\frac{\partial}{\partial y_1}(h_2A_2) - \frac{\partial}{\partial y_2}(h_1A_1)\right]\right\}. \tag{V–40}$$

The Laplacian is given by

$$\nabla^2\phi = \frac{1}{h_1h_2h_3}\sum_{s=1}^{3}\frac{\partial}{\partial y_s}\left(h_1h_2h_3\frac{1}{h_s^2}\frac{\partial\phi}{\partial y_s}\right). \tag{V–41}$$

The previous identities still hold, that is,

$$\begin{aligned} \nabla\times\nabla\phi &= 0,\\ \nabla\cdot(\nabla\times\mathbf{A}) &= 0\\ \nabla\times(\nabla\times\mathbf{A}) &= \nabla(\nabla\cdot\mathbf{A}) - \nabla^2\mathbf{A}, \end{aligned} \tag{V–42}$$

as do also the Gauss and Stokes theorems. Also, an arbitrary vector **A** can always be written

$$\mathbf{A} = \nabla\phi + \nabla \times \mathbf{B}, \tag{V–43}$$

where $\nabla \cdot \mathbf{B}$ is arbitrary. The material derivative of any quantity is still given by

$$\frac{d\phi}{dt} = \frac{\partial\phi}{\partial t} + \nabla\phi \cdot \mathbf{v}. \tag{V–44}$$

Consider next the equations of elasticity. Define σ_{ij} as the component in the y_j-parametric direction of the stress vector acting on an area element perpendicular to the y_i-parametric direction. Clearly, $\sigma_{ij} = \sigma_{ji}$. The equations of equilibrium become:

$$\sum_{m=1}^{3} \left\{ \frac{\partial}{\partial y_m}\left(\frac{h_i}{h_m}\sigma_{im}\right) + \frac{1}{h_1h_2h_3}\left[\frac{\partial}{\partial y_m}(h_1h_2h_3)\right]\sigma_{im}\frac{h_i}{h_m} - \sigma_{mm}\frac{1}{h_m}\frac{\partial h_m}{\partial y_i} \right\} + \rho P_i h_i = 0, \tag{V–45}$$

where P_i is the body force component in the ith parametric direction.[8] The strain tensor ϵ_{ij} is defined similarly with respect to the curvilinear system in terms of elongation and angle change; again $\epsilon_{ij} = \epsilon_{ji}$, and

$$\epsilon_{ij} = \frac{1}{2}\left[\frac{h_i}{h_j}\frac{\partial}{\partial y_j}\left(\frac{u_i}{h_i}\right) + \frac{h_j}{h_i}\frac{\partial}{\partial y_i}\left(\frac{u_j}{h_j}\right) + 2\delta_{ij}\sum_{s=1}^{3}\frac{u_s}{h_sh_j}\frac{\partial h_i}{\partial y_s}\right], \tag{V–46}$$

where u_i is the displacement-vector component in the y_i-parametric direction.

The relation between σ_{ij} and ϵ_{ij} is exactly Eqs. (V–11) and (V–12), so that the equivalent of the Navier equations may be obtained by substitution of Eq. (V–46) into Eq. (V–45).

The compatibility equations involve second derivatives and so are more complicated. Define

$$\Gamma_{ij}{}^{m} = \frac{1}{2h_m{}^2}\left(\delta_{im}\frac{\partial h_i{}^2}{\partial y_j} + \delta_{jm}\frac{\partial h_j{}^2}{\partial y_i} - \delta_{ij}\frac{\partial h_j{}^2}{\partial y_m}\right),$$

$$C_{ijk} = \frac{\partial}{\partial y_k}(h_ih_j\sigma_{ij}) - \sum_{m=1}^{3}(\Gamma_{ik}{}^{m}h_mh_j\sigma_{mj} + \Gamma_{jk}{}^{m}h_mh_i\sigma_{mi}).$$

[8] Remember that the summation convention of Chapter I is not being used in Sec. (V–6).

Then the stress-compatibility equations, for the case of a body force per unit mass of P_i in the ith parametric direction, become:

$$\sum_{j=1}^{3} \frac{1}{h_j^2}\left(C_{ikj;j} + \frac{1}{1+\sigma} C_{jji;k}\right) + \frac{\partial}{\partial y_k}(\rho h_i P_i) + \frac{\partial}{\partial y_i}(\rho h_k P_k)$$

$$+ \sum_{m=1}^{3} \left\{(-2\Gamma_{ik}{}^m \rho h_m P_m) + \frac{\sigma h_i^2}{1-\sigma}\delta_{ik}\left[\frac{1}{h_m^2}\frac{\partial}{\partial y_m}(\rho h_m P_m)\right.\right.$$

$$\left.\left. - \sum_{s=1}^{3}\left(\frac{1}{h_s^2}\Gamma_{ss}{}^m \rho h_m P_m\right)\right]\right\} = 0, \qquad \text{(V-47)}$$

where the semicolon operation is defined by

$$C_{ijk;s} = \frac{\partial}{\partial y_s} C_{ijk} - \sum_{m=1}^{3} (\Gamma_{is}{}^m C_{mjk} + \Gamma_{js}{}^m C_{imk} + \Gamma_{ks}{}^m C_{ijm}).$$

V-7. Anisotropy

Anisotropy was discussed briefly in Sec. V-1; we now return to it and establish the form of the stress-strain law for this more general case.

Consider an elastic body whose properties depend on direction, so that, for example, the response to a tensile load in one direction will differ from the response to the same tensile load applied in another direction. Then any stress-strain law that is derived can in general hold for only one specific orientation of the body, and so it will be assumed that this orientation has been prescribed once and for all. If it is again assumed, as in Sec. V-1, that the relation between stress and strain is linear, we can write

$$\tau_{ij} = c_{ijkl} e_{kl}. \qquad \text{(V-48)}$$

Since $\tau_{ij} = \tau_{ji}$, it follows that the constants c_{ijkl} are symmetric in their first two subscripts, so that $c_{ijkl} = c_{jikl}$. There is no loss of generality in also requiring symmetry in the second pair of subscripts, for if there happened to be a lack of such symmetry it would only be necessary to replace each e_{ij} by $\frac{1}{2}(e_{ij} + e_{ji})$.

It is more surprising that there should be symmetry between the two pairs of indices, in the sense that $c_{ijkl} = c_{klij}$. This result follows from energy considerations. The rate at which mechanical work is being done on an elastic body at any instant is

$$\begin{aligned} \dot{W} &= \int \rho F_i \dot{u}_i \, dV + \int T_i \dot{u}_i \, dS \\ &= \int (\rho F_i \dot{u}_i + \tau_{ij,j} \dot{u}_i + \tau_{ij} \dot{u}_{i,j}) \, dV \\ &= \int \rho \ddot{u}_i \dot{u}_i \, dV + \int \tau_{ij} \tfrac{1}{2} (\dot{u}_{i,j} + \dot{u}_{j,i}) \, dV \\ &= \dot{K} + \int \tau_{ij} \dot{e}_{ij} \, dV, \end{aligned}$$

where $\dot{K}$ is the rate of accumulation of kinetic energy, given by

$$\begin{aligned} \dot{K} &= \frac{d}{dt} \int \frac{1}{2} \rho \dot{u}_i \dot{u}_i \, dV \\ &= \int \rho \ddot{u}_i \dot{u}_i \, dV \end{aligned}$$

(cf. Eq. (IV–8)), and where in writing

$$\dot{e}_{ij} = \tfrac{1}{2} (\dot{u}_{i,j} + \dot{u}_{j,i})$$

nonlinear contributions have of course been neglected, in accordance with the lack of distinction between Eulerian and Lagrangian coordinates in this chapter. It follows that the rate of increase of mechanical strain energy of the body (again, thermal effects are considered minor, a fact which will be subsequently verified by the more exact analysis of Chapters VIII and X) is given by

$$\frac{d}{dt} \int U \rho \, dV = \dot{W} - \dot{K} = \int \tau_{ij} \dot{e}_{ij} \, dV,$$

where U is the strain energy per unit mass of the material. This relation must hold for arbitrary volumes of integration, so that

$$\rho \dot{U} = \tau_{ij} \dot{e}_{ij}.$$

If now U, a function of the e_{ij}, is written symmetrically in terms of the e_{ij} (which can always be done by replacing e_{ij} by $\frac{1}{2}(e_{ij} + e_{ji})$ if necessary), this result becomes

$$\rho \frac{\partial U}{\partial e_{ij}} \dot{e}_{ij} = \tau_{ij} \dot{e}_{ij},$$

and it can be concluded because of the symmetry that

$$\tau_{ij} = \rho \frac{\partial U}{\partial e_{ij}}. \qquad \text{(V–49)}$$

It is incidentally immaterial whether ρ is evaluated at the instant in question or at zero time, for the net effect of any discrepancy would be of second order. Expanding U in a power series gives

$$U = \alpha + \beta_{ij}e_{ij} + \gamma_{ijkl}e_{ij}e_{kl} + \cdots, \tag{V–50}$$

where α is unimportant and where β_{ij} must vanish if zero stress is to correspond to zero strain. Then Eq. (V–49) gives

$$\begin{aligned} \tau_{ij} &= \rho(\gamma_{ijkl} + \gamma_{klij})e_{kl} \\ &= c_{ijkl}e_{kl}, \end{aligned}$$

and clearly

$$c_{ijkl} = c_{klij},$$

which is the desired result. It is possible to insist that $\gamma_{ijkl} = \gamma_{klij}$, so that the elastic energy is given by

$$\rho U = \tfrac{1}{2}\tau_{ij}e_{ij}.$$

The various symmetries in the subscripts are easily seen to imply that at most 21 of the c_{ijkl} can differ from one another in numerical value. (At one time it was argued by Cauchy and Poisson on the basis of a certain molecular model that the number could be reduced to 15, but this result was found not to agree with experiment.) If there are symmetries in a crystal, then it may be expected that the physical properties exhibit the same symmetries, and so the number of independent constants can be reduced. A survey of crystal geometric and physical properties has been given by Voigt [9] and summarized by Love.[10] There is no difficulty in modifying the previous elasticity equations so as to hold for any particular anisotropic body.

[9] W. Voigt, *Lehrbuch der Krystallphysik* (Teubner, Berlin, 1910).

[10] A. E. H. Love, *Mathematical theory of elasticity* (4th ed.; Dover, New York, 1945) p. 149.

VI

General Solutions

VI–1. Review of Potential Theory

Not only are the equations of elasticity analogous to those of potentia theory — as has previously been remarked — but the techniques of potential theory may in fact be used to obtain general solutions in elasticity It is therefore worth while to review briefly the applicable portions o potential theory.

The most useful tool is the divergence theorem, which states (in eithe two or three dimensions) that

$$\int \phi_{i,i}\, dV = \int \phi_i n_i\, dS.$$

Given any two functions A, B, we may define a vector ϕ_i by $\phi_i = AB_{,i}$ the divergence theorem applied to this ϕ_i gives

$$\int AB_{,ii}\, dV + \int A_{,i}B_{,i}\, dV = \int A\,\frac{\partial B}{\partial n}\, dS, \qquad \text{(VI–1)}$$

which is Green's first identity. Interchanging A and B and subtractin the new result from the previous one gives Green's second identity:

$$\int (AB_{,ii} - BA_{,ii})\, dV = \int \left(A\,\frac{\partial B}{\partial n} - B\,\frac{\partial A}{\partial n}\right) dS. \qquad \text{(VI–2)}$$

By setting $A = 1$ in Eq. (VI–1), and letting B be harmonic, it is seen that

$$\int \frac{\partial B}{\partial n}\, dS = 0 \qquad \text{(VI–3)}$$

for any harmonic function. A similar situation arises in elasticity, wher the surface integral of the stress vector must vanish (for zero body force)

For two harmonic functions A and B, Eq. (VI–2) is analogous to the eciprocal theorem of elasticity.

Let (ξ_i) be a fixed interior point of the region V, and define $r(x;\xi)$ as he distance between any other point (x_i) and (ξ_i); that is,

$$r(x;\xi) = [(x_1 - \xi_1)^2 + (x_2 - \xi_2)^2 + (x_3 - \xi_3)^2]^{\frac{1}{2}}.$$

Then as long as (x_i) does not coincide with (ξ_i), the function $1/r(x;\xi)$ is asily seen to be harmonic. Consider a region V' which is identical with V except that a small sphere of radius R and surrounding (ξ_i) has been deleted. Equation (VI–2) with $B = 1/r(x;\xi)$ may be applied to this deeted region, provided that we remember there are now two bounding urfaces. On the spherical surface, the outward unit normal from V' has omponents $(\xi_i - x_i)/R$, and also

$$\left(\frac{1}{r}\right)_{,i} = -\frac{x_i - \xi_i}{R^3}.$$

The result is

$$-\int \frac{\nabla^2 A}{r}\, dV' = \int \left[A \frac{\partial}{\partial n}\left(\frac{1}{r}\right) - \frac{1}{r}\frac{\partial A}{\partial n}\right] dS_1 + \int \left[A \frac{1}{R^2} - \frac{1}{R} A_{,i} \frac{\xi_i - x_i}{R}\right] dS_0,$$

where S_1 is the common exterior surface of V and V', and where S_0 is the surface of the small sphere surrounding (ξ_i). As $R \to 0$, this result learly becomes

$$4\pi A(\xi) = -\int \frac{\nabla^2 A}{r(x;\xi)}\, dV + \int \left[\frac{1}{r(x;\xi)}\frac{\partial A}{\partial n} - A \frac{\partial}{\partial n}\left(\frac{1}{r(x;\xi)}\right)\right] dS, \quad \text{(VI–4)}$$

where (ξ_i) is held fixed in the integration. This is Green's third identity, nd holds for an arbitrary (but of course sufficiently well-behaved) function A. If A is harmonic, then only the second integral on the right-hand side remains; it then follows that the value of A at any point ξ can be etermined if the boundary values of A and $\partial A/\partial n$ are prescribed. The various physical analogies from fluid mechanics, electrostatics, and so on, imply, however, that giving only one of A or $\partial A/\partial n$ on the boundary should be sufficient. In the second case, the condition (VI–3) must be atisfied. The fluid analogy — that the net outflow of fluid be zero — shows that this condition is physically both necessary and sufficient.

Define therefore a function $v(x;\xi)$ to be harmonic in x and to equal $-1/r(x;\xi)$ on the surface. Equation (VI–2) requires that

$$0 = -\int v(x;\xi) \cdot \nabla^2 A\, dV - \int \left(A \frac{\partial v}{\partial n} + \frac{1}{r}\frac{\partial A}{\partial n}\right) dS,$$

which when added to Eq. (VI–4) gives

$$4\pi A(\xi) = -\int G(x;\xi) \cdot \nabla^2 A \, dV - \int A \frac{\partial G(x;\xi)}{\partial n} \, dS, \qquad \text{(VI–5)}$$

where the integration is with respect to x, for fixed ξ, and where the Green's function G is defined by

$$G(x;\xi) = v(x;\xi) + \frac{1}{r(x;\xi)}.$$

Thus if G is known (and in principle the point-source physical interpretation of G guarantees its existence), and if $\nabla^2 A$ is given (zero for Laplace's equation; some prescribed function of position in Poisson's equation), Eq. (VI–5) is a formal solution of the Dirichlet problem in which A is prescribed on the boundary.

It may be noted that (1), the basic function $1/r$ used above is the so-called fundamental solution of Laplace's equation in three dimensions, in the sense that it is the only nontrivial solution depending only on r (in two dimensions, the fundamental solution in r would be used in a precisely similar manner); (2), G is symmetric so that $G(\alpha;\beta) = G(\beta;\alpha)$ as is easily shown by writing Eq. (VI–2) for $G(x;\alpha)$ and $G(x;\beta)$ and for a region doubly deleted by removal of small spheres surrounding (α) and (β); (3), if the Laplacian with respect to ξ is taken of both sides in Eq. (VI–4), then, since $\nabla^2(1/r) = \nabla^2(1/r)_{,i} = 0$, the result is

$$4\pi \nabla^2 A = -\int \nabla^2 A \cdot \nabla^2(1/r) \, dV,$$

and since $\nabla^2 A$ is arbitrary this implies that

$$\int f \nabla^2(1/r) \, dV = -4\pi f(\xi)$$

for any function f, that is, the quantity $\nabla^2(1/r)$ acts like an operator (Dirac delta function type) which annihilates f except at $x = \xi$, where it gives a concentrated contribution of $-4\pi f$.

If secondly $\partial A/\partial n$ is prescribed on the boundary, with

$$\int \frac{\partial A}{\partial n} \, dS = 0, \qquad \text{(VI–6)}$$

then a "Green's function of the second kind" may be used to solve this Neumann problem. Note first that setting $A = 1$ in Eq. (VI–4) gives

$$4\pi = -\int \frac{\partial}{\partial n} \frac{1}{r(x;\xi)} \, dS. \qquad \text{(VI–7)}$$

efine $h(x;\xi)$, for fixed ξ, to be harmonic in x, and to satisfy the boundary ondition

$$\frac{\partial h}{\partial n} = -\frac{\partial}{\partial n}\left(\frac{1}{r}\right) - \frac{4\pi}{S_0},$$

here S_0 is the total surface area. By Eq. (VI–7), this boundary condi-on is permissible, and so h exists. Equation (VI–2) gives

$$0 = -\int h\nabla^2 A \, dV - \int \left(A\frac{\partial h}{\partial n} - h\frac{\partial A}{\partial n}\right) dS,$$

hich when added to Eq. (VI–4) gives

$$4\pi A(\xi) = -\int H(x;\xi)\cdot\nabla^2 A \, dV + \frac{4\pi}{S}\int A \, dS + \int H(x;\xi)\frac{\partial A}{\partial n}\, dS, \tag{VI–8}$$

here

$$H(x;\xi) = h(x;\xi) + \frac{1}{r(x;\xi)},$$

nd where the integration is with respect to x for fixed ξ. The second erm on the right-hand side is a constant, which may be disregarded ince the solution to the Neumann problem is clearly indeterminate ithin a constant anyway. Note that for the same reason $h(x;\xi)$ and so $(x;\xi)$ are indeterminate within a function of ξ; it is easily shown by the me method used for G that this arbitrary function of ξ may be so chosen at $H(x;\xi)$ is symmetric.

It may be remarked that the uniqueness proofs for the Dirichlet and eumann problems (Sec. I–13, Ex. 3) are entirely analogous to that for e uniqueness theorem of elasticity (Sec. V–4).

I–2. Somigliana's Integrals

In Sec. (VI–1), the fundamental solution $1/r$ was inserted into Green's cond identity to obtain Green's third identity, Eq. (VI–4). It was re-arked that the second identity and the elastic reciprocity theorem are nalogous; also, Eq. (V–22) gives a fundamental elasticity solution which rresponds to $1/r$. Consequently we are led to apply the reciprocal eorem with $u_i^{(j)}(x;\xi)$ as the displacement resulting from a concentrated nit body force applied in the x_j-direction and at the point (ξ_1, ξ_2, ξ_3).

Consider therefore an elasticity problem in which it is desired to deter-ine u_i in the interior of the body when certain boundary conditions e prescribed. Let T_i be the surface stresses (which may not be known), and let F_i be the body force. A second elasticity solution for the sam region is given by the displacement $u_i^{(j)}(x;\xi)$ (variable x_i, fixed ξ_i), whic can be maintained by the application of a concentrated unit force at and by appropriate surface tractions $T_i^{(j)}$. The reciprocal theorem ap plied to these two solutions (treating the concentrated force as a limi of finite forces applied over a continually decreasing volume) gives

$$\int T_i u_i^{(j)}\, dS + \int \rho F_i u_i^{(j)}\, dV = \int T_i^{(j)} u_i\, dS + u_j(\xi), \qquad \text{(VI–9}$$

where the integrations are with respect to x, ξ being held fixed. The anal ogy to Eq. (VI–4) is evident, particularly in view of Eq. (V–20), whic relates T_i to $\partial u_i/\partial n$.

Suppose now that only u_i is prescribed on the boundary. Proceedin in the same general way as in the derivation of Eq. (VI–5), define $v_i^{(}$ to be a displacement solution of the elasticity equations for $F_i = 0$ sat isfying the boundary condition

$$v_i^{(j)} = -u_i^{(j)}.$$

The reciprocal theorem gives

$$\int T_i v_i^{(j)}\, dS + \int \rho F_i v_i^{(j)}\, dV = \int Q_i^{(j)}\, u_i\, dS,$$

where $Q_i^{(j)}$ is the surface traction required for $v_i^{(j)}$. Adding to Eq. (VI–9 gives (cf. Eq. (VI–5)):

$$u_j(\xi) = -\int G_i^{(j)}(x;\xi) u_i\, dS + \int \rho F_i g_i^{(j)}(x;\xi)\, dV, \qquad \text{(VI–10}$$

where

$$G_i^{(j)}(x;\xi) = T_i^{(j)}(x;\xi) + Q_i^{(j)}(x;\xi)$$

and

$$g_i^{(j)}(x;\xi) = u_i^{(j)}(x;\xi) + v_i^{(j)}(x;\xi),$$

so that $G_i^{(j)}$ are the surface tractions corresponding to the displacemen $g_i^{(j)}$. In Eq. (VI–10), ξ is a fixed parameter and integration is with respec to x. Physical considerations guarantee the existence of $g_i^{(j)}$ and $G_i^{(j)}$ so that Eq. (VI–10) then gives a formal solution of the elasticity problen in which u_i is prescribed on the boundary. Note that $g_i^{(j)}(x;\xi)$ is the dis placement at x due to a unit concentrated j-direction force at ξ, wher the boundary is held fixed. This interpretation, together with the recip rocal theorem, provides an immediate proof of the symmetry property

$$g_i^{(j)}(x;\xi) = g_j^{(i)}(\xi;x).$$

As a matter of fact, the definition of $g_i^{(j)}$ could have been used directly in the reciprocal theorem so as to give Eq. (VI–10) without the intermediate step of Eq. (VI–9); the present method was used in order to emphasize the analogy with potential theory.

In potential theory, the Green's function for a sphere with the pole at the origin is trivial; an immediate consequence of Eq. (VI–5) is then that the value of a harmonic function at any point is equal to the average of its values on a sphere surrounding that point. There is an analogous result in elasticity (due to Aquaro and Synge); it may first be checked that, for a sphere of radius R,

$$g_i^{(j)}(x;0) = \frac{1}{4\pi E}\left(\frac{1+\sigma}{1-\sigma}\right)\left\{\left(\frac{3}{2} - 2\sigma\right)\delta_{ij}\left(\frac{1}{r} - \frac{1}{R}\right) + \frac{1}{2}\frac{x_i x_j}{r^3}\right.$$
$$\left. - \frac{1}{2R^3}\left[-\frac{3-2\sigma}{4-6\sigma}(r^2 - R^2)\delta_{ij} + x_i x_j\right]\right\}, \quad \text{(VI–11)}$$

for this displacement vector vanishes at the boundary and satisfies the Navier equations everywhere except at the origin, where by virtue of Eq. (V–22) it corresponds to a unit force in the x_j-direction. Calculation of $G_i^{(j)}$ and use of Eq. (VI–10), for the case of zero body force, gives

$$u_j(0) = -\frac{3\sigma - \frac{3}{4}}{4\pi R^2(2-3\sigma)}\int u_j\, dS + \frac{15}{16\pi R^2(2-3\sigma)}\int u_r n_j\, dS, \quad \text{(VI–12)}$$

where u_r is the outward radial component of the displacement on the surface of the sphere. If therefore the displacements on the surface of any sphere drawn in the material are known, the displacement at the center of the sphere may be calculated by this mean-value theorem. For the commonly occurring case $\sigma = \frac{1}{4}$, the result becomes

$$u_j(0) = \frac{3}{4\pi R^2}\int u_r n_j\, dS, \quad \text{(VI–13)}$$

and it is interesting to note that in this case only radial and not tangential displacement has an effect on $u_j(0)$. Equation (VI–13) may also be derived by writing the Navier equations in the form

$$u_i + \frac{1}{2(1-2\sigma)}\, x_i u_{j,j} = \text{harmonic function},$$

setting $\sigma = \frac{1}{4}$, and using the result that the value of a harmonic function at the center of a sphere is equal to its average over the volume of the sphere. This same method may also be used, although with more labor to derive Eq. (VI–12).

It has been pointed out by Diaz and Payne[1] that the biharmonic nature of u_i and τ_{ij} may be used as a basis for deriving various mean value theorems of elasticity. If ϕ is a biharmonic function, then

$$\phi(0) = \frac{15}{8\pi R^3}\int \phi\, dV - \frac{3}{8\pi R^2}\int \phi\, dS,$$

so that each of u_i, τ_{ij} satisfies this equation. Manipulation of integrals with use of Eq. (V–13) or (V–14) then gives the desired results. Some of these results are:

$$e_{ik}(0) = \frac{3}{8\pi R^3(7-10\sigma)}\Big[-10\sigma\int (u_i n_k + u_k n_i)\, dS + 35(1-2\sigma)\int n_i n_k n_j u_j\, dS - 7\delta_{ik}\int u_j n_j\, dS\Big]$$

from which $\tau_{ik}(0)$ may be obtained by use of Hooke's law, and (Synge)

$$\tau_{ij}(0) = \frac{3}{8\pi R^2(7+5\sigma)}\Big[10\sigma\int T_i n_j\, dS - 7\delta_{ij}\int T_k n_k\, dS + 35\int T_k n_k n_i n_j\, dS\Big]$$

(Note that $\int T_i n_j\, dS = \int T_j n_i\, dS$.)

Return now to Eq. (VI–9), and suppose secondly that the tractions T are prescribed on the boundary. The tractions $T_i{}^{(j)}$ are in equilibrium with a unit force at ξ, so that

$$\int T_i{}^{(j)}\, dS + \delta_{ij} = 0,$$

$$\int e_{isk} x_s T_k{}^{(j)}\, dS + e_{isk}\xi_s\delta_{jk} = 0.$$

Consider therefore a new problem in which the tractions are defined by

$$Q_i{}^{(j)} = -T_i{}^{(j)} + \frac{1}{V}\big[\tfrac{1}{2}(\xi_i n_j - \delta_{ij}\xi_s n_s) - x_i n_j\big], \qquad \text{(VI–14}$$

[1] J. B. Diaz and L. E. Payne, "Mean value theorems in the theory of elasticity," *Proceedings of the Third United States Congress of Applied Mechanics* (American Society of Mechanical Engineers, New York, 1958).

where V is the volume of the body. By use of the divergence theorem, it is easily checked that these surface stresses satisfy

$$Q_i^{(j)}\,dS = \int e_{isk}x_sQ_k^{(j)}\,dS = 0,$$

so that an associated displacement vector $v_i^{(j)}(x;\xi)$ corresponding to zero body force does indeed exist. The reciprocal theorem requires that

$$\int T_iv_i^{(j)}\,dS + \int \rho F_iv_i^{(j)}\,dS = \int Q_i^{(j)}u_i\,dS,$$

and adding this result to Eq. (VI–9) gives, with Eq. (VI–14),

$$\int T_ih_i^{(j)}(x;\xi)\,dS + \int \rho F_ih_i^{(j)}(x;\xi)\,dV$$
$$= \frac{1}{V}\int\left[\frac{1}{2}(\xi_in_j - \delta_{ij}\xi_sn_s) - x_in_j\right]u_i\,dS + u_j(\xi), \quad \text{(VI–15)}$$

where ξ is fixed during the integration, and where

$$h_i^{(j)}(x;\xi) = u_i^{(j)}(x;\xi) + v_i^{(j)}(x;\xi).$$

The first term on the right-hand side of Eq. (VI–15) is unknown since u_i is not prescribed on the boundary; however, this term is a linear function of (ξ_i) with skew-symmetric coefficients and so merely represents rigid-body motion, which is arbitrary anyway. Consequently it may be disregarded, and Eq. (VI–15) is then the desired expression for u_i at an interior point in terms of surface and body forces.

Physically, $h_i^{(j)}(x;\xi)$ is the displacement corresponding to a concentrated force (in the x_j-direction) at ξ, balanced by a surface stress vector

$$(1/V)[\tfrac{1}{2}(\xi_in_j - \delta_{ij}\xi_sn_s) - x_in_j].$$

Since a prescription of surface and body stresses leaves the amount of rigid-body motion undefined, $h_i^{(j)}(x;\xi)$ is indeterminate within a function of the form

$$[f_{is}^{(j)}(\xi)]x_s + g_i^{(j)}(\xi),$$

where $f_{is}^{(j)}$ is skew-symmetric in its two subscripts. For suitable choice of these arbitrary functions, the reader may easily show by use of the reciprocal theorem that the resultant $h_i^{(j)}$ is symmetric in the sense that

$$h_i^{(j)}(x;\xi) = h_j^{(i)}(\xi;x).$$

Once the functions $g_i^{(j)}$ and $h_i^{(j)}$ have been determined for a certain elastic body, Eqs. (VI–10) and (VI–15) may be used to write down the solution corresponding to any pure-displacement or pure-traction boundary condition. The method is convenient if a number of boundary conditions are to be analyzed for the same body, or if it is easier to find the functions $g_i^{(j)}$ or $h_i^{(j)}$ than it would be to solve the desired problem directly (note the relatively simple boundary conditions satisfied by these special functions). Less often, it is useful to have a similar expression for the mixed boundary-value problem in which displacements are specified on a portion S_u of the boundary and tractions on a portion S_τ of the boundary. Define the displacement field $p_i^{(j)}(x;\xi)$ as that which corresponds to a unit concentrated force at (ξ) in the (x_j) direction; the associated surface traction $P_i^{(j)}$ is to vanish on S_τ and $p_i^{(j)}$ is to vanish on S_u. Then the reciprocal theorem gives

$$u_j(\xi) = -\int P_i^{(j)} u_i \, dS_u + \int \rho F_i p_i^{(j)} \, dV + \int T_i p_i^{(j)} \, dS_\tau,$$

where the integration is with respect to x, with fixed ξ. The same kind of symmetry condition as before is satisfied here also:

$$p_i^{(j)}(x;\xi) = p_j^{(i)}(\xi;x).$$

The equations of this section provide a formal solution for the displacement vector u_i in terms of integrals involving the prescribed surface forces and displacements and body forces. It need hardly be remarked that other quantities of interest, such as e_{ij}, ω_{ij}, and τ_{ij}, can then be obtained by appropriate differentiation. The equations for two of the quantities, $u_{i,i}$ and ω_{ij}, will be obtained by a slightly different method in the next section.

VI–3. Betti's Integrals

Suppose first that the displacement vector u_i is prescribed on the boundary, and that in some way $u_{i,i}$ has been determined from these boundary conditions. Then Eq. (V–13) states that the Laplacian of u is the known function

$$-\frac{1}{1-2\sigma} u_{j,ji} - \frac{1}{G} \rho F_i,$$

and so it is only necessary to solve a Poisson equation of the type discussed in Sec. VI–1. The situation is particularly simple if the body force is zero, for then the function ϕ_i defined by

$$\phi_i = u_i + \frac{1}{2(1-2\sigma)} x_i u_{s,s}$$

is harmonic (cf. Eq. (V–19)) and has known boundary values, so that the problem reduces to the standard Dirichlet type.

If secondly the stress vector T_i is prescribed, then Eq. (V–20) implies that the boundary values of $\partial u_i/\partial n$ are known if ω_{ij} and $u_{k,k}$ are known; consequently, if ω_{ij} is known at least on the boundary and $u_{k,k}$ everywhere, then the problem again reduces via Eq. (V–13) to a Poisson type.

The quantities $u_{i,i}$ and ω_{ij} may be determined formally from surface data by methods entirely similar to those of Secs. VI–1 and VI–2. Consider first $u_{i,i}$. Denote by $u_i^{(0)}(x;\xi)$ the displacement vector

$$\frac{\partial}{\partial x_i}\left(\frac{1}{r(x;\xi)}\right),$$

where

$$r(x;\xi) = [(x_1 - \xi_1)^2 + (x_2 - \xi_2)^2 + (x_3 - \xi_3)^2]^{\frac{1}{2}},$$

and note that $u_i^{(0)}(x;\xi)$ is a displacement vector of the concentrated expansive-force type discussed in Sec. V–3. Applying the reciprocal theorem to the body as modified by the removal of a small spherical region surrounding ξ, and noting that in the limit a contribution at ξ is received from both sides of the theorem, the result is easily seen to be

$$\frac{8\pi G(1-\sigma)}{1-2\sigma}\, u_{i,i}(\xi) = \int u_i T_i^{(0)}\, dS - \int u_i^{(0)} T_i\, dS - \int u_i^{(0)} \rho F_i\, dV. \tag{VI–16}$$

In this equation (the analog of Green's third identity), ξ is held fixed during the integration with respect to x.

If the displacements are prescribed on the boundary, we utilize an auxiliary displacement function which coincides with $u_i^{(0)}$ on the boundary; if T_i is prescribed, we choose an auxiliary displacement which gives rise to surface tractions coinciding with $T_i^{(0)}$ (which are in equilibrium). In either case, the combination of the auxiliary displacement and $u_i^{(0)}$ produces a Green's function in a way which is by now familiar, so that the problem is formally solved.

In the case of zero body force, two sphere theorems for the dilatation are easily derived from the fact that $u_{i,i}$ is harmonic so that the mean-value theorem requires

$$\begin{aligned} u_{i,i}(0) &= \frac{3}{4\pi R^3}\int u_{i,i}\, dV \\ &= \frac{3}{4R\pi^3}\int u_i n_i\, dS = \frac{3}{4\pi R^3}\int u_r\, dS, \end{aligned}$$

where u_r is the radial component of displacement. Alternatively, the first integral may be written

$$u_{i,i}(0) = \frac{3}{4\pi R^3}\frac{1-2\sigma}{E}\int \tau_{ii}\,dV$$

$$= \frac{3}{4\pi R^3}\frac{1-2\sigma}{E}\int (\tau_{ij}x_j)_{,i}\,dV$$

$$= \frac{3}{4\pi R^2}\frac{1-2\sigma}{E}\int T_r\,dS,$$

where T_r is the radial component of surface traction.

Consider now ω_{12}. Define $u_i^{(M)}(x;\xi)$ by

$$u_i^{(M)}(x;\xi) = e_{3it}\left(\frac{1}{r}\right)_{,t},$$

which by Eq. (V–28) corresponds to a concentrated couple at ξ. The reciprocal theorem yields

$$8\pi G\omega_{21}(\xi) = \int T_i u_i^{(M)}\,dS - \int T_i^{(M)} u_i\,dS + \int \rho F_i u_i^{(M)}\,dV, \quad \text{(VI–17)}$$

and Green's functions may be constructed as before to give a formal solution for ω_{21} when either displacements or tractions are specified on the surface. The other components of rotation are handled similarly.

As in the previous section, there is no difficulty in extending the Betti method to the case of mixed boundary conditions; the details may be left to the reader.

VI–4. Solutions in Terms of Special Functions

It will be assumed in the following that body forces have been removed by the method of Sec. V–3. Then Eqs. (V–13) imply that each displacement component is biharmonic, which in turn implies that each u_i can be expressed in terms of harmonic functions.

For let ϕ be any biharmonic function, and define a harmonic function ψ such that $\psi_{,1} = \nabla^2\phi$. This can always be done, for the Laplacian of any ψ satisfying $\psi_{,1} = \nabla^2\phi$ must be a function of (x_2, x_3) only, say $f(x_2, x_3)$, and so if a function $\Omega(x_2, x_3)$ satisfying $\Omega_{,22} + \Omega_{,33} = -f$ (Poisson's equation, which by physical considerations of electrostatics and other examples always has a solution) is added to ψ, then $\psi_{,1}$ is unaffected and $\psi_{,ii}$ is now zero. Then write

$$\phi = \beta + \tfrac{1}{2}x_1\psi$$

and it follows that β must be harmonic.

(a) Betti representation

Consequently each u_i could be expressed in terms of two harmonic functions in this way, a total of six functions being needed. However, the requisite number of functions may be reduced by using Eq. (V–16) [2] which states that the curl of $u_{i,ss}$ is zero, so that $u_{i,ss}$ must be the gradient of some scalar function 2α:

$$u_{i,ss} = 2\alpha_{,i},$$

and since $u_{i,iss} = 0$ by Eq. (V–17), it follows that α is harmonic. Write

$$u_i = \gamma_i + x_i\alpha \qquad \text{(VI–18)}$$

and then γ_i must be harmonic because of the definition of α. Thus the three u_i may always be expressed in terms of four harmonic functions γ_i, α; however, not every choice of harmonic (γ_i, α) satisfies Eqs. (V–13), so that an auxiliary requirement is that

$$(5 - 4\sigma)\alpha + x_j\alpha_{,j} + \gamma_{j,j} = \text{constant}. \qquad \text{(VI–19)}$$

For the case $\sigma = \frac{1}{2}$, this constant must be zero. For a given u_i, α is determined within a constant by its definition, and so γ_i is determined by Eq. (VI–18) within a constant times x_i.

(b) Kelvin's first representation

Let F be a harmonic function satisfying $F_{,1} = \alpha$, where α is defined as above; the existence of such an F is guaranteed by the reasoning of the introduction to this section. Then the quantity $x_i\alpha - x_1F_{,i}$ is harmonic, so that a representation equivalent to that of Eq. (VI–18) is

$$u_i = H_i + x_1F_{,i}, \qquad \text{(VI–20)}$$

where the four functions H_i, F are harmonic. Such a representation is always possible; however, not every choice of harmonic H_i, F yields a solution because of the requirement from Eq. (V–13) that

$$(3 - 4\sigma)F_{,1} + H_{j,j} = \text{constant}, \qquad \text{(VI–21)}$$

[2] This is preferable to using the Navier equations, which do not hold for $\sigma = \frac{1}{2}$.

where for $\sigma = \frac{1}{2}$ the constant must be zero. By an appropriate modification in the definition of F, either x_2 or x_3 could of course be used instead of x_1.

(c) *Boussinesq representation*

Let u_i be any elastic displacement vector. Then since α and therefore $F_{,1}$ contains an arbitrary constant, a set of harmonic (F, H_i) can always be found such that, by Eqs. (VI–20) and (VI–21),

$$u_i = H_i + x_1 F_{,i}, \tag{VI–22}$$

$$(3 - 4\sigma)F_{,1} + H_{j,j} = 0. \tag{VI–23}$$

Define a set of three harmonic functions J_i by

$$J_{i,1} = H_i$$

(again, see the introduction of this section for proof that this is possible), and note that each J_i is still arbitrary within a harmonic $g_i(x_2, x_3)$. Now Eq. (VI–23) requires that

$$(3 - 4\sigma)F + J_{i,i} = f(x_2, x_3),$$

where f is harmonic, so choose the g_i such that this f is equal to zero (for example, g_2 alone need differ from zero to do this). Thus there always exists a set of three harmonic functions J_i such that

$$u_i = J_{i,1} - \frac{x_1}{3 - 4\sigma} J_{j,ji}, \tag{VI–24}$$

and moreover any set of harmonic J_i generates a solution of the Navier equations (V–13). Equation (VI–24) is also valid for the case $\sigma = \frac{1}{2}$.

(d) *Kelvin's second representation*

Define α as in (a), and define β to be a harmonic solution of

$$\beta + 2x_p\beta_{,p} = \alpha. \tag{VI–25}$$

It is first necessary to show that such a solution exists, and this will be done here only for a region in which every boundary point can be reached by an uninterrupted straight line drawn from the origin (such a region, whose entire boundary surface would be directly illuminated by a point light source placed at the origin, is said to be "star-shaped").

For this kind of region, spherical coordinates (r, θ, ϕ) are appropriate;

Eq. (VI–25) becomes

$$\beta + 2r\frac{\partial\beta}{\partial r} = \alpha,$$

and it is easily checked by direct substitution that a well-behaved solution of this equation is

$$\beta(r, \theta, \phi) = \frac{1}{\sqrt{r}}\int_0^r \frac{\alpha(t, \theta, \phi)}{2\sqrt{t}}\,dt.$$

Note that $\beta(0, \theta, \phi) = \alpha(0, \theta, \phi)$ as required by Eq. (VI–25). Applying the Laplacian operator

$$\begin{aligned}\nabla^2 &= \frac{1}{r^2}\left[\frac{\partial}{\partial r}\left(r^2\frac{\partial}{\partial r}\right) + \frac{1}{\sin\theta}\frac{\partial}{\partial\theta}\left(\sin\theta\frac{\partial}{\partial\theta}\right) + \frac{1}{\sin^2\theta}\frac{\partial^2}{\partial\phi^2}\right]\\ &= \frac{1}{r^2}\left[\frac{\partial}{\partial r}\left(r^2\frac{\partial}{\partial r}\right) + L_{\theta,\phi}\right]\end{aligned}$$

to β gives

$$\nabla^2\beta = \frac{1}{2r^2}\left(\frac{1}{2}\alpha + r\frac{\partial\alpha}{\partial r} - \frac{1}{4\sqrt{r}}\int_0^r\frac{\alpha}{\sqrt{t}}\,dt + \frac{1}{\sqrt{r}}\int_0^r\frac{L\alpha}{\sqrt{t}}\,dt\right).$$

Writing (because of the harmonicity of α)

$$L\alpha = -\frac{\partial}{\partial t}\left(t^2\frac{\partial\alpha}{\partial t}\right)$$

in the last integral and integrating by parts shows that

$$\nabla^2\beta = 0,$$

which is the required result.

Define now the functions ζ_i by

$$u_i = \zeta_i + r^2\beta_{,i}, \tag{VI–26}$$

where $r^2 = x_jx_j$. Then use of Eq. (VI–25) and the definition of α shows that each ζ_i is harmonic. Thus any elastic-displacement vector corresponding to zero body force may be expressed in terms of four harmonic functions and r^2, provided that the body is star-shaped. An alternative form of Eq. (VI–26) which is convenient for elastic-sphere (radius a) problems is

$$u_i = \zeta_i + (r^2 - a^2)\beta_{,i}.$$

Although every displacement may be expressed in the form of Eq. (VI–26), it is not true that any set of (β, ζ_i) will generate a solution corresponding to zero body force, for Eq. (V–13) requires that

$$(2 - 4\sigma)\beta + (6 - 8\sigma)x_s\beta_{,s} + \zeta_{j,j} = \text{constant}, \qquad \text{(VI–27)}$$

where the constant for the case of incompressibility must be taken as zero.

(*e*) *Papkovich-Neuber representation*

Let θ be any function whose Laplacian equals $u_{i,i}$. Write

$$u_i = \theta_{,i} + \lambda_i;$$

then $\lambda_{i,i} = 0$. Equation (V–13) then requires that

$$u_{i,jj} + \frac{1}{1 - 2\sigma}\,\theta_{,jji} = 0,$$

so that there exist harmonic functions η_i such that

$$u_i = -\frac{1}{1 - 2\sigma}\,\theta_{,i} + \eta_i. \qquad \text{(VI–28)}$$

Taking the divergence of Eq. (VI–28) gives

$$\theta_{,ii} = -\frac{1}{1 - 2\sigma}\,\theta_{,ii} + \eta_{i,i},$$

so that

$$\theta\left(\frac{2 - 2\sigma}{1 - 2\sigma}\right) = \frac{1}{2}\,x_j\eta_j - (2 - 2\sigma)\omega,$$

where ω is harmonic. Consequently Eq. (VI–28) becomes

$$u_i = -\frac{1}{4 - 4\sigma}\,(x_j\eta_j)_{,i} + \eta_i + \omega_{,i}, \qquad \text{(VI–29)}$$

which is therefore a representation for u_i in terms of four harmonic functions. Further, any choice of (η_i, ω) satisfies the Navier equations.

In the derivation of this result, Navier's equation (V–13) was used, and since this equation is not valid for $\sigma = \frac{1}{2}$, the incompressible case requires special treatment. For $\sigma = \frac{1}{2}$, define ν such that

$$\nu_{,jj} + 2\alpha = 0,$$

where α is defined in section (*a*) above. Then define

$$\eta_i = u_i + \nu_{,i}$$

and note that η_i is then harmonic, and also that η_i and u_i have the same curl. Define μ_i by

$$u_i = -\tfrac{1}{2}(x_j\eta_j)_{,i} + \eta_i + \mu_i.$$

Taking the curl of both sides shows that the curl of μ_i vanishes, so that μ_i can be written as $\omega_{,i}$; taking the divergence of the equation then shows that ω is harmonic, so Eq. (VI–29) is obtained here also. Again, any choice of harmonic (η_i, ω) satisfies Eqs. (V–16) and also $u_{i,i} = 0$.

For star-shaped regions at least (see Sec. VI–4(d)), the number of harmonic functions required in Eq. (VI–29) may be reduced by one. For define a harmonic function p such that

$$4(1-\sigma)p - x_j p_{,j} = -4(1-\sigma)\omega$$

and, by the argument of section (d), such a p will exist except possibly when 4σ is an integer. Define

$$\eta_i = \eta_i' + p_{,i},$$

and then Eq. (VI–29) gives

$$u_i = -\frac{1}{4(1-\sigma)}(x_j\eta_j')_{,i} + \eta_i', \tag{VI–30}$$

so that in Eq. (VI–29) the function ω may be omitted without loss of generality provided that 4σ is not an integer. (An example given by Eubanks and Sternberg [3] shows that if 4σ is an integer then the reduction need not be possible.)

Secondly, define μ' to be a harmonic solution of $\mu_{,1}' = \eta_1'$, and define $\eta_i'' = \eta_i' - \mu_{,i}'$, so that $\eta_1'' = 0$. Then Eq. (VI–30) gives

$$u_i = \eta_i'' - \frac{1}{4(1-\sigma)}(x_j\eta_j'')_{,i} + \left[\mu' - \frac{1}{4(1-\sigma)}(x_j\mu_{,j}')\right]_{,i}, \tag{VI–31}$$

where the quantity in brackets is harmonic; Eq. (VI–31) has the form of Eq. (VI–29), with the feature that $\eta_1'' = 0$, so that again a reduction to three harmonic functions has been made.

(f) *Galerkin representation*

Since any vector may be written as the sum of an irrotational and a solenoidal part, write

$$u_i = (1-2\sigma)L_{,i} - 2(1-\sigma)e_{ijk}M_{k,j},$$

[3] R. A. Eubanks and E. Sternberg, "On the completeness of the Boussinesq-Papkovich stress functions," *J. Rational Mech. and Analysis 5* (1956), 735.

where (Sec. I–13) $M_{k,k}$ can be set equal to zero without loss of generality. This representation holds also for $\sigma = \frac{1}{2}$, for it then becomes equivalent to the condition $u_{i,i} = 0$. Since $M_{k,k} = 0$,

$$M_k = e_{kst} P_{t,s}$$

where $P_{t,t}$ will be chosen equal to L. Then

$$u_i = -P_{t,ti} + 2(1-\sigma) P_{i,jj} \qquad \text{(VI–31)}$$

and Eq. (V–13) requires each P_i to be biharmonic. Also, any biharmoni[c] (P_i) generates an elasticity solution. For $\sigma = \frac{1}{2}$, the condition $u_{i,i} =$ [0] is automatically satisfied, and Eq. (V–17) requires that the curl of (P_i) be biharmonic; however, adding the gradient of a suitable function t[o] (P_i) shows that the condition of biharmonicity is sufficiently general her[e] also. For $\sigma = \frac{1}{2}$, an alternative form of Eq. (VI–31) is

$$\mathbf{u} = -\nabla \times (\nabla \times \mathbf{P}).$$

(*g*) *Maxwell-Morera representation*

In any of the previous representations, the components of the stress tensor may be calculated from the displacements so as to give an expression for τ_{ij} in terms of harmonic or biharmonic functions — and moreover one which automatically satisfies the equations of compatibility. In cases (*c*), (*e*), and (*f*), the equations of equilibrium are automatically satisfied; in the other cases, the various auxiliary conditions must hold. However, the resulting expressions are complicated, and so it is often convenient to use general solutions derived specifically for use with stresses, and such are the Maxwell and Morera solutions.

The Maxwell solution is obtained by defining three functions ϕ_{11}, ϕ_{22}, ϕ_{33} such that

$$\tau_{12} = -\phi_{33,12}, \quad \tau_{13} = -\phi_{22,13}, \quad \tau_{23} = -\phi_{11,23}. \qquad \text{(VI–32)}$$

Then the equilibrium equations (V–7) require that, within certain arbitrary functions of integration,

$$\begin{aligned} \tau_{11} &= \phi_{22,33} + \phi_{33,22}, \\ \tau_{22} &= \phi_{33,11} + \phi_{11,33}, \\ \tau_{33} &= \phi_{11,22} + \phi_{22,11}. \end{aligned} \qquad \text{(VI–33)}$$

The arbitrariness in the ϕ-functions defined by Eq. (VI–32) is easily seen to be sufficient that the above-mentioned functions of integration can be omitted, and it will be assumed that this has been done. Thus any stress state may be expressed in terms of three functions by means o[f]

Eqs. (VI–32) and (VI–33), and, conversely, any choice of these three functions is such that the corresponding stresses satisfy the equations of equilibrium.

The set of ϕ-functions is not uniquely determined by Eqs. (VI–32) and (VI–33); for subsequent use in simplifying the compatibility equations, it is necessary to consider next the degree of indeterminacy in the ϕ-functions. If the stresses are all zero, then Eqs. (VI–32) require that

$$\begin{aligned} \phi_{11}{}^0 &= f_1(x_1, x_2) + g_1(x_1, x_3), \\ \phi_{22}{}^0 &= f_2(x_2, x_3) + g_2(x_2, x_1), \\ \phi_{33}{}^0 &= f_3(x_3, x_1) + g_3(x_3, x_2), \end{aligned}$$

where f_i and g_i are as yet arbitrary functions. The first of Eqs. (VI–33) gives

$$f_{2,33} + g_{3,22} = 0. \tag{VI–34}$$

Define $G_1{}^0(x_2, x_3)$ to be any function such that

$$G_{1,33}{}^0 = g_3.$$

Then Eq. (VI–34) gives

$$f_2 + G_{1,22}{}^0 = x_3 A(x_2) + B(x_2).$$

Define

$$G_1 = G_1{}^0 - x_3 C(x_2) - D(x_2),$$

where $C_{,22} = A$ and $D_{,22} = B$; then g_3 is also equal to $G_{1,33}$, and

$$f_2 + G_{1,22} = 0.$$

The other equations in (VI–33) give similar results, so that the most general ϕ-functions corresponding to zero stress are:

$$\begin{aligned} \phi_{11}{}^0 &= [G_2(x_1, x_3) - G_3(x_1, x_2)]_{,11}, \\ \phi_{22}{}^0 &= [G_3(x_1, x_2) - G_1(x_2, x_3)]_{,22}, \\ \phi_{33}{}^0 &= [G_1(x_2, x_3) - G_2(x_1, x_3)]_{,33}. \end{aligned} \tag{VI–35}$$

Consequently the ϕ-functions defined by Eqs. (VI–32) and (VI–33) are undetermined within the ϕ^0-functions of Eqs. (VI–35), where the G_i functions are arbitrary. The equations of compatibility (V–14) require that (using the fact that τ_{ss} is harmonic)

$$\begin{aligned} &-[(1+\sigma)\nabla^2\phi_{33} - \tau_{ss}]_{,12} = 0, \\ &-[(1+\sigma)\nabla^2\phi_{22} - \tau_{ss}]_{,13} = 0, \\ &-[(1+\sigma)\nabla^2\phi_{11} - \tau_{ss}]_{,23} = 0, \\ &[(1+\sigma)\nabla^2\phi_{22} - \tau_{ss}]_{,33} + [(1+\sigma)\nabla^2\phi_{33} - \tau_{ss}]_{,22} = 0, \\ &[(1+\sigma)\nabla^2\phi_{33} - \tau_{ss}]_{,11} + [(1+\sigma)\nabla^2\phi_{11} - \tau_{ss}]_{,33} = 0, \\ &[(1+\sigma)\nabla^2\phi_{11} - \tau_{ss}]_{,22} + [(1+\sigma)\nabla^2\phi_{22} - \tau_{ss}]_{,11} = 0. \end{aligned}$$

These equations have exactly the same form as those used to determine the ϕ^0 functions, and so the solution must have the same form as Eqs. (VI–35), involving now arbitrary functions H_1, H_2, H_3. Incorporate therefore in the ϕ-functions a set of G-functions such that

$$-(1+\sigma)\nabla^2 G_i = H_i,$$

and then the compatibility equations can be written

$$\nabla^2\phi_{11} = \nabla^2\phi_{22} = \nabla^2\phi_{33} = \frac{\tau_{ss}}{1+\sigma}. \qquad \text{(VI–36)}$$

Using Eq. (VI–33), this may be rewritten

$$\nabla^2\phi_{11} = \nabla^2\phi_{22} = \nabla^2\phi_{33} = \frac{\phi_{11,11} + \phi_{22,22} + \phi_{33,33}}{2-\sigma}. \qquad \text{(VI–37)}$$

To summarize: it has been shown that there always exists a set of three stress functions satisfying the equilibrium equations (V–7) and the compatibility equations (VI–36), in terms of which the stresses may be expressed by means of Eqs. (VI–32) and (VI–36).

The first of Eqs. (VI–36) may be written

$$\phi_{11,11} + (\tau_{33} - \phi_{22,11}) + (\tau_{22} - \phi_{33,11}) = \frac{\tau_{ss}}{1+\sigma},$$

which is equivalent to

$$(1+\sigma)(\phi_{11,1} - \phi_{22,1} - \phi_{33,1})_{,1} = Ee_{11}.$$

If the displacement equations obtained in this way are integrated, then it is found that, apart from a rigid-body motion,

$$u_1 = \frac{1+\sigma}{E}(\phi_{11} - \phi_{22} - \phi_{33})_{,1},$$

$$u_2 = \frac{1+\sigma}{E}(\phi_{22} - \phi_{11} - \phi_{33})_{,2},$$

$$u_3 = \frac{1+\sigma}{E}(\phi_{33} - \phi_{11} - \phi_{22})_{,3}.$$

The Morera stress functions are defined in a manner complementary to that of Eqs. (VI–32) and (VI–33):

$$\begin{aligned}
\tau_{11} &= -2\phi_{23,23},\\
\tau_{22} &= -2\phi_{31,31},\\
\tau_{33} &= -2\phi_{12,12},\\
\tau_{12} &= (\phi_{23,1} + \phi_{31,2} - \phi_{12,3})_{,3},\\
\tau_{13} &= (\phi_{12,3} + \phi_{23,1} - \phi_{31,2})_{,2},\\
\tau_{23} &= (\phi_{31,2} + \phi_{12,3} - \phi_{23,1})_{,1},
\end{aligned} \qquad \text{(VI–38)}$$

and again there is enough arbitrariness in the ϕ-functions that this definition is legitimate. Any choice of ϕ-functions in Eqs. (VI–38) will satisfy the equations of equilibrium. The ϕ-functions must of course satisfy the compatibility conditions (V–14) if they are to represent an elasticity solution; these conditions do not reduce as simply as in the case of the Maxwell functions.

If the Maxwell and Morera representations are added, the result may be written (defining ϕ_{ij} to be symmetric):

$$\tau_{ij} = e_{ipq}e_{jmr}\phi_{qr,pm}. \tag{VI–39}$$

Thus, if such a set of quantities ϕ_{ij} has been found for one coordinate system, and if ϕ_{ij}' for another system are calculated by the rules of Cartesian tensor transformation applied to ϕ_{ij}, Eq. (VI–39) will hold in all coordinate systems. An equation of the form of (VI–39) could have been obtained directly from the equilibrium equations by noting that $\tau_{ij,j} = 0$ implies that there exists a set of quantities A_{ij} such that

$$\tau_{ij} = e_{jst}A_{it,s}$$

and then treating the requirement $\tau_{ij} = \tau_{ji}$ similarly. It may also be noted that the relation between the quantities defined by Eqs. (VI–32) and (VI–33) and those defined by Eqs. (VI–38) is exactly the same (except for a sign) as that between the e_{ij} as expressed in the compatibility equations (IV–34).

VI–5. Decomposition of Stress and Strain Tensors

For zero body force, the equations of equilibrium are of the "divergence-free" type; also, the equations of compatibility are the conditions that a "potential," namely, the displacement vector, should exist, and so are analogous to the irrotationality condition in vector analysis. Since an arbitrary vector field can be written as the sum of a solenoidal and an irrotational field, one might anticipate that a similar result holds for an arbitrary second-order symmetric tensor, the decomposition now being in terms of two symmetric tensors, one of which satisfies equilibrium and the other of which satisfies compatibility.

Consider an elastic body occupying a volume V with surface S. Let the body force be F_i, and let the boundary conditions be such that any one portion of the boundary falls into one of the following classes (the subscripts n and t refer to normal and tangential components respectively):

(*a*) u_n and u_t prescribed,
(*b*) u_n and T_t prescribed,
(*c*) u_t and T_n prescribed,
(*d*) T_n and T_t prescribed.

In the special case when (*d*) holds over the entire surface, over-all equilibrium must of course be satisfied. Let σ_{ij} be an arbitrary symmetric tensor, which we think of as a mythical stress state.

Define τ_{ij}'' to be the stress state solving the following elastic problem:

$$\tau_{ij,j}'' = (\sigma_{ij,j} + \rho F_i),$$
$$Ee_{ij}'' = (1 + \sigma)\tau_{ij}'' - \sigma\delta_{ij}\tau_{kk}'',$$
$$e_{ij}'' = \tfrac{1}{2}(u_{i,j}'' + u_{j,i}''),$$

where moreover u_i'' is to agree with all prescribed boundary displacement conditions, and where the stress vector $\sigma_{ij}n_j - \tau_{ij}''n_j$ is to agree with all prescribed stress boundary conditions. Define next τ_{ij}' by

$$\sigma_{ij} = \tau_{ij}' + \tau_{ij}''. \tag{VI–40}$$

Then

$$\tau_{ij,j}' = -\rho F_i,$$

and $T_i' = \tau_{ij}'n_j$ agrees with all prescribed stress boundary conditions.

Thus Eq. (VI–40) provides a decomposition of the arbitrary symmetric tensor σ_{ij} into two stress states, one of which satisfies equilibrium and the other of which satisfies compatibility (not the Beltrami-Mitchell form, however, for the equilibrium equations were used in that derivation); moreover, the two stress states satisfy the stress and displacement boundary conditions, respectively.

It is clear that this result could have been framed in terms of arbitrary symmetric "strain" tensors instead of arbitrary "stress" tensors σ_{ij}. There is also no difficulty in extending the theorem to hold for more complicated types of boundary conditions.

VI–6. Function Space Methods

Consider an elastic body under the action of a certain body force and with prescribed boundary conditions of the type described in Sec. VI–5. It is desired to find a stress state which satisfies the equations of equilibrium and compatibility, and such that the resultant surface tractions and displacements satisfy the boundary conditions. Let **S** denote a completely arbitrary stress state, which need satisfy none of these conditions; to specify **S**, the six components of the stress tensor must be stated at every point of the body, and so in a sense **S** may be thought of as a vector with an infinite number of components. This generalization of the idea of vectors is fruitful only if other concepts, such as magnitude, scalar product, and so forth, can also be carried over, and for the case of elasticity this is indeed possible. Consider two arbitrary stress states, $\mathbf{S}_1$ and $\mathbf{S}_2$. Let the stresses and strains (these latter calculated from Hooke's law) be denoted by $(\tau_{ij}^{(1)}, e_{ij}^{(1)})$, $(\tau_{ij}^{(2)}, e_{ij}^{(2)})$ respectively. Define the scalar product by

$$\mathbf{S}_1 \cdot \mathbf{S}_2 = \int \tau_{ij}^{(1)} e_{ij}^{(2)}\, dV = \int \tau_{ij}^{(2)} e_{ij}^{(1)}\, dV = \mathbf{S}_2 \cdot \mathbf{S}_1$$

and then the magnitude of $\mathbf{S}_1$ is given by

$$| \mathbf{S}_1 |^2 = \mathbf{S}_1 \cdot \mathbf{S}_1,$$

which is always positive, vanishing only when $\mathbf{S}_1 = 0$. The result of multiplying **S** by a constant k is defined to be a new stress state each of whose stress components is k times the previous one; thus $(\mathbf{S}/| \mathbf{S} |)$ is a unit vector. The equation

$$(\mathbf{S}_1 + x\mathbf{S}_2) \cdot (\mathbf{S}_1 + x\mathbf{S}_2) = 0 \qquad \text{(VI–41)}$$

where x is a scalar, may be expanded by use of the distributive law (easily checked to be valid for generalized vectors) to give

$$\mathbf{S}_1 \cdot \mathbf{S}_1 + 2x(\mathbf{S}_1 \cdot \mathbf{S}_2) + x^2(\mathbf{S}_2 \cdot \mathbf{S}_2) = 0. \qquad \text{(VI–42)}$$

Now because of the positive definite character of the scalar product, Eq. (VI–41) can have no real solutions for x (unless S_1 and S_2 are proportional, in which case there is exactly one real solution), so that the discriminant of Eq. (VI–42) must be nonpositive (vanishing only in the case of proportionality). Consequently,

$$(\mathbf{S}_1 \cdot \mathbf{S}_2)^2 \leq (\mathbf{S}_1 \cdot \mathbf{S}_1)(\mathbf{S}_2 \cdot \mathbf{S}_2), \qquad \text{(VI–43)}$$

where equality holds only in the case of proportionality. This is *Schwartz's inequality*.

An immediate result of Eq. (VI–43) is that the *triangle inequality* holds for these generalized vectors; that is,

$$| (| \mathbf{S}_1 | - | \mathbf{S}_2 |) | \leq | \mathbf{S}_1 + \mathbf{S}_2 | \leq | \mathbf{S}_1 | + | \mathbf{S}_2 |. \qquad \text{(VI–44)}$$

Also, the angle θ between S_1 and S_2 may now be defined by

$$\cos\theta = \frac{\mathbf{S}_1 \cdot \mathbf{S}_2}{|\mathbf{S}_1| \cdot |\mathbf{S}_2|}, \tag{VI–45}$$

and Eq. (VI–43) ensures that θ is real. If $\cos\theta = 0$, then $\mathbf{S}_1$ and $\mathbf{S}_2$ are said to be *orthogonal.*

Let $\mathbf{S}_1$, $\mathbf{S}_2$, $\mathbf{S}_3$ be a set of three arbitrary vectors. Then it is always possible to find a set of no more than three unit vectors, $\mathbf{I}_1$, $\mathbf{I}_2$, $\mathbf{I}_3$, mutually orthogonal, such that each $\mathbf{S}_i$ is a linear combination of the $\mathbf{I}_i$. This *Gram-Schmid* procedure is as follows:

(*a*) define $\mathbf{I}_1 = \dfrac{\mathbf{S}_1}{|\mathbf{S}_1|}$;

(*b*) define $\mathbf{I}_2 = \dfrac{\mathbf{S}_2 - (\mathbf{S}_2 \cdot \mathbf{I}_1)\mathbf{I}_1}{|\mathbf{S}_2 - (\mathbf{S}_2 \cdot \mathbf{I}_1)\mathbf{I}_1|}$;

(*c*) define $\mathbf{I}_3 = \dfrac{\mathbf{S}_3 - (\mathbf{S}_3 \cdot \mathbf{I}_1)\mathbf{I}_1 - (\mathbf{S}_3 \cdot \mathbf{I}_2)\mathbf{I}_2}{|\mathbf{S}_3 - (\mathbf{S}_3 \cdot \mathbf{I}_1)\mathbf{I}_1 - (\mathbf{S}_3 \cdot \mathbf{I}_2)\mathbf{I}_2|}$;

and it is easily checked that the result is the desired one. Exactly the same procedure may be used for any number of vectors $\mathbf{S}_i$.

Let $(\mathbf{I}_1, \mathbf{I}_2, \cdots, \mathbf{I}_n)$ be any set of n such *orthonormal* vectors. Then the generalized *Pythagoras theorem* states that, for any arbitrary $\mathbf{S}$,

$$|\mathbf{S}|^2 = |\mathbf{S} - \sum_{i=1}^{n} (\mathbf{S} \cdot \mathbf{I}_i)\mathbf{I}_i|^2 + \sum_{i=1}^{n} (\mathbf{S} \cdot \mathbf{I}_i)^2, \tag{VI–46}$$

and this is easily proved by simply expanding the first term on the right-hand side. From Eq. (VI–46) follows *Bessel's inequality,*

$$(\mathbf{S} \cdot \mathbf{S}) \geq \sum_{i=1}^{n} (\mathbf{S} \cdot \mathbf{I}_i)^2. \tag{VI–47}$$

The quantities $(\mathbf{S} \cdot \mathbf{I}_i)$ may be thought of as the components of $\mathbf{S}$ along the unit vectors $\mathbf{I}_i$, and the foregoing results then have a simple geometric interpretation for vectors $\mathbf{S}$ thought of as emanating from the origin of an infinite-dimensional space.

Returning now to the elasticity problem, function space methods for delineating the range within which the solution vector must lie have been given by Prager and Synge [4] and by Diaz.[5]

[4] W. Prager and J. L. Synge, "Approximations in elasticity based on the concept of function space," *Quart. Appl. Math. 5* (1947), 241; *6* (1948), *15.*

[5] J. B. Diaz, *Inequalities and minimal principles in mathematical physics* (Lecture Series, No. 18; University of Maryland Institute for Fluid Dynamics and Applied Mathematics, College Park, 1956).

Define **S** to be the solution vector, and let **S**′ denote a stress state which satisfies the equation of equilibrium,

$$\tau_{ij,j}' + \rho F_i = 0,$$

but not necessarily compatibility; in addition, **S**′ must satisfy all boundary conditions on stress. Let $\mathbf{I}_p'$ ($p = 1, 2, \cdots, m$) be any set of orthonormal stress states, each of which satisfies equilibrium with zero body force, but not necessarily compatibility; further, each $\mathbf{I}_p'$ must provide a surface stress which vanishes wherever surface traction is prescribed (thus if T_t is prescribed over part of the surface, we require $\tau_{ij}'^{(p)} n_j$ to have no tangential component over that part of the surface). Define **S**″ to be a stress state satisfying the equations of compatibility (Eqs. (V–10) rather than (V–14), since these latter involve equilibrium), but not necessarily the equations of equilibrium. Because compatibility is satisfied, displacements u_i'' exists; the displacements are required to satisfy the boundary conditions wherever displacement is prescribed. Finally, define $\mathbf{I}_q''$ ($q = 1, 2, \cdots, n$) to be a set of orthonormal stress states each of which satisfies compatibility (Eqs. (V–10) in terms of stresses) but not necessarily equilibrium; the corresponding displacements must vanish wherever surface displacements are prescribed.

Consider now the quantity

$$\begin{aligned}(\mathbf{S} - \mathbf{S}') \cdot (\mathbf{S} - \mathbf{S}'') &= \int (\tau_{ij} - \tau_{ij}')(e_{ij} - e_{ij}'')\, dV \\ &= \int (\tau_{ij} - \tau_{ij}')(u_{i,j} - u_{i,j}'')\, dV \\ &= \int [(\tau_{ij} - \tau_{ij}')(u_i - u_i'')]_{,j}\, dV,\end{aligned}$$

because of the fact that $\tau_{ij,j} = \tau_{ij,j}' = -\rho F_i$. Continuing, the right-hand side becomes, by use of the divergence theorem,

$$\int (T_i - T_i')(u_i - u_i'')\, dS,$$

which must vanish because of the nature of the boundary conditions. Thus

$$\mathbf{S} \cdot \mathbf{S} - \mathbf{S}' \cdot \mathbf{S} - \mathbf{S}'' \cdot \mathbf{S} + \mathbf{S}' \cdot \mathbf{S}'' = 0, \tag{VI–48}$$

that is,

$$|\,\mathbf{S} - \tfrac{1}{2}(\mathbf{S}' + \mathbf{S}'')\,|^2 = \tfrac{1}{4}|\,\mathbf{S}' - \mathbf{S}''\,|^2, \tag{VI–49}$$

which states that the end point of the vector **S** must lie on a hypersphere whose center is at $\frac{1}{2}(\mathbf{S}' + \mathbf{S}'')$ and which has radius $\frac{1}{2}|\,\mathbf{S}' - \mathbf{S}''\,|$. The possible geometric configuration of the desired solution is therefore known.

In Eq. (VI–48), $\mathbf{S}' + k'\mathbf{I}_p'$ may be substituted for $\mathbf{S}'$ and $\mathbf{S}'' + k''\mathbf{I}_q''$ for $\mathbf{S}''$; the result must hold for arbitrary constants k', k'' and so

$$\mathbf{S} \cdot \mathbf{I}_p' = \mathbf{S}'' \cdot \mathbf{I}_p', \tag{VI–50}$$

$$\mathbf{S} \cdot \mathbf{I}_q'' = \mathbf{S}' \cdot \mathbf{I}_q'', \tag{VI–51}$$

$$\mathbf{I}_p' \cdot \mathbf{I}_q'' = 0. \tag{VI–52}$$

Equation (VI–50) states that the projection of $\mathbf{S}$ onto each $\mathbf{I}_p'$ is the same as that of $\mathbf{S}''$, and so each $\mathbf{I}_p'$ determines a hyperplane perpendicular to it, in which the terminus of $\mathbf{S}$ must lie. Equation (VI–51) has a similar interpretation, and so Eqs. (VI–50) and (VI–51) require that $\mathbf{S}$ must lie in a certain hyperplane which is itself the intersection of a number of hyperplanes. The intersection of this hyperplane with the hypersphere is a hypercircle, in which the terminus of $\mathbf{S}$ must therefore lie.

Analytically, we replace $\mathbf{S}'$ by $\mathbf{S}' + \sum k_p'\mathbf{I}_p'$ and $\mathbf{S}''$ by $\mathbf{S}'' + \sum k_q''\mathbf{I}_q''$ in Eq. (VI–49), and then choose the k_p' and the k_q'' so as to minimize the radius of the sphere. The result is easily seen to be that the extremity of $\mathbf{S}$ lies on a hypercircle of center

$$\mathbf{C} = \tfrac{1}{2}\{\mathbf{S}' + \mathbf{S}'' - \sum \mathbf{I}_p'[\mathbf{I}_p' \cdot (\mathbf{S}' - \mathbf{S}'')] + \sum \mathbf{I}_q''[\mathbf{I}_q'' \cdot (\mathbf{S}' - \mathbf{S}'')]\} \tag{VI–53}$$

and of radius R given by

$$R^2 = \tfrac{1}{4}\{(\mathbf{S}' - \mathbf{S}'') \cdot (\mathbf{S}' - \mathbf{S}'') - \sum [(\mathbf{S}' - \mathbf{S}'') \cdot \mathbf{I}_p']^2 - \sum [(\mathbf{S}' - \mathbf{S}'') \cdot \mathbf{I}_q'']^2\}. \tag{VI–54}$$

If it is possible to find a complete set of $\mathbf{I}_p'$, $\mathbf{I}_q''$ such that $\mathbf{S}' - \mathbf{S}''$ can be expressed as a linear combination of them, then R of Eq. (VI–54) would be zero and so an exact solution $\mathbf{S} = \mathbf{C}$ would be obtained.

It is of course also possible to attempt a direct expansion of $\mathbf{S}$ in terms of a complete set of orthonormal states $\mathbf{J}_i$. For simplicity, let $F_i = 0$; let each $\mathbf{J}_i$ satisfy equilibrium with zero body force and also compatibility (such $\mathbf{J}_i$ may, for example, be obtained by use of the general solutions of Sec. VI–4). Then writing

$$\mathbf{S} = \sum \mathbf{J}_i(\mathbf{S} \cdot \mathbf{J}_i)$$

gives the solution, and this form is useful if the quantities $(\mathbf{S} \cdot \mathbf{J}_i)$ can be evaluated from the known boundary conditions — as can, for example, be done if either stresses or displacements are everywhere specified.

It should also be remarked that in the various results above it is just as easy to center attention on displacement states instead of stress states.

There are a number of useful inequalities which can be written down. From Eq. (VI–47),

$$\mathbf{S}\cdot\mathbf{S} \geq \sum (\mathbf{I}_p'\cdot\mathbf{S})^2 + \sum (\mathbf{I}_q''\cdot\mathbf{S})^2 \geq \sum (\mathbf{I}_p'\cdot\mathbf{S}'')^2 + \sum (\mathbf{I}_q''\cdot\mathbf{S}')^2. \tag{VI–55}$$

An upper bound can be obtained by noting that Eq. (VI–49) is equivalent to

$$\mathbf{S} = \tfrac{1}{2}(\mathbf{S}' + \mathbf{S}'') + \tfrac{1}{2}|\,\mathbf{S}' - \mathbf{S}''\,|\mathbf{J},$$

where $\mathbf{J}$ is a unit vector, so that by the triangle inequality (VI–44)

$$|\,\mathbf{S}\,| \leq \tfrac{1}{2}|\,\mathbf{S}' + \mathbf{S}''\,| + \tfrac{1}{2}|\,\mathbf{S}' - \mathbf{S}''\,|.$$

Squaring both sides,

$$\begin{aligned}\mathbf{S}\cdot\mathbf{S} &\leq \tfrac{1}{2}(\mathbf{S}'\cdot\mathbf{S}' + \mathbf{S}''\cdot\mathbf{S}'') + \tfrac{1}{2}|\,\mathbf{S}' + \mathbf{S}''\,|\cdot|\,\mathbf{S}' - \mathbf{S}''\,| \\ &\leq \tfrac{1}{2}(\mathbf{S}'\cdot\mathbf{S}' + \mathbf{S}''\cdot\mathbf{S}'') + \tfrac{1}{2}[(\mathbf{S}'\cdot\mathbf{S}' + \mathbf{S}''\cdot\mathbf{S}'')^2 - 4(\mathbf{S}'\cdot\mathbf{S}'')^2]^{\frac{1}{2}}.\end{aligned} \tag{VI–56}$$

The weaker result

$$\mathbf{S}\cdot\mathbf{S} \leq \mathbf{S}'\cdot\mathbf{S}' + \mathbf{S}''\cdot\mathbf{S}'' \tag{VI–57}$$

is rather interesting. Note that in Eqs. (VI–56) and (VI–57), $\mathbf{S}'$ may be replaced by $\mathbf{S}' - \sum (\mathbf{S}'\cdot\mathbf{I}_p')\mathbf{I}_p'$, and $\mathbf{S}''$ by $\mathbf{S}'' - \sum (\mathbf{S}''\cdot\mathbf{I}_q'')\mathbf{I}_q''$, so as to strengthen the bound. Finally, it follows easily from Eq. (VI–48) that

$$\begin{aligned}|\,\mathbf{S} - \mathbf{S}'\,| &\leq |\,\mathbf{S}' - \mathbf{S}''\,|, \\ |\,\mathbf{S} - \mathbf{S}''\,| &\leq |\,\mathbf{S}' - \mathbf{S}''\,|.\end{aligned} \tag{VI–58}$$

Alternative inequalities are obtained by making use of the previous hypercircle interpretation. From Eq. (VI–49), $\mathbf{S}$ is known to lie on a hypercircle of radius R_1 and with center $\mathbf{C}_1$, given by $R_1 = \frac{1}{2}\,|\,\mathbf{S}' - \mathbf{S}''\,|$, $\mathbf{C}_1 = \frac{1}{2}(\mathbf{S}' + \mathbf{S}'')$. Also, the combined set $\{\mathbf{I}_p', \mathbf{I}_q''\}$ forms one large orthonormal set (cf. Eq. (VI–52)) which will be denoted by $\{\mathbf{I}_i\}$. Then

$$\begin{aligned}|\,\mathbf{S}\,|^2 &= \sum (\mathbf{S}\cdot\mathbf{I}_i)^2 + |\,\mathbf{S} - \sum (\mathbf{S}\cdot\mathbf{I}_i)\mathbf{I}_i\,|^2 \qquad \text{(Pythagoras)} \\ &= \sum (\mathbf{S}\cdot\mathbf{I}_i)^2 + \\ &\qquad |\,\mathbf{S} - \sum (\mathbf{S}\cdot\mathbf{I}_i)\mathbf{I}_i - \{\mathbf{C}_1 - \sum (\mathbf{C}_1\cdot\mathbf{I}_i)\mathbf{I}_i\} + \{\mathbf{C}_1 - \sum (\mathbf{C}_1\cdot\mathbf{I}_i)\mathbf{I}_i\}\,|^2 \\ &= \sum (\mathbf{S}\cdot\mathbf{I}_i)^2 + |\,\mathbf{S} - \mathbf{C}_1 - \sum ((\mathbf{S} - \mathbf{C}_1)\cdot\mathbf{I}_i)\mathbf{I}_i\,|^2 + |\,\mathbf{C}_1 - \sum (\mathbf{C}_1\cdot\mathbf{I}_i)\mathbf{I}_i\,|^2 \\ &\qquad + 2(\mathbf{S} - \mathbf{C}_1 - \sum ((\mathbf{S} - \mathbf{C}_1)\cdot\mathbf{I}_i)\mathbf{I}_i)\cdot(\mathbf{C}_1 - \sum (\mathbf{C}_1\cdot\mathbf{I}_i)\mathbf{I}_i).\end{aligned}$$

Applying the Schwartz inequality to the last term, that is,

$$-|\boldsymbol{\alpha}|\cdot|\boldsymbol{\beta}| \leq \boldsymbol{\alpha}\cdot\boldsymbol{\beta} \leq |\boldsymbol{\alpha}|\cdot|\boldsymbol{\beta}|,$$

and using the fact that

$$|\mathbf{S}-\mathbf{C}_1| = R_1,$$

we obtain the following double inequality:

$$\mathbf{S}\cdot\mathbf{S} \leq \sum(\mathbf{S}\cdot\mathbf{I}_i)^2 + \{[\mathbf{C}_1\cdot\mathbf{C}_1 - \sum(\mathbf{C}_1\cdot\mathbf{I}_i)^2]^{\frac{1}{2}} + [R_1^2 - \sum(\mathbf{S}\cdot\mathbf{I}_i - \mathbf{C}_1\cdot\mathbf{I}_i)^2]^{\frac{1}{2}}\}^2,$$

$$\mathbf{S}\cdot\mathbf{S} \geq \sum(\mathbf{S}\cdot\mathbf{I}_i)^2 + \{[\mathbf{C}_1\cdot\mathbf{C}_1 - \sum(\mathbf{C}_1\cdot\mathbf{I}_i)^2]^{\frac{1}{2}} - [R_1^2 - \sum(\mathbf{S}\cdot\mathbf{I}_i - \mathbf{C}_1\cdot\mathbf{I}_i)^2]^{\frac{1}{2}}\}^2,$$

which may be simplified to read

$$\mathbf{C}\cdot\mathbf{C} + R^2 - 2R\,|\mathbf{G}| < \mathbf{S}\cdot\mathbf{S} \leq \mathbf{C}\cdot\mathbf{C} + \mathrm{R}^2 + 2R\,|\mathbf{G}|,$$

where $\mathbf{C}$ and R are given by Eqs. (VI–53) and (VI–54), and where

$$|\mathbf{G}|^2 = \tfrac{1}{4}\{(\mathbf{S}'+\mathbf{S}'')(\mathbf{S}'+\mathbf{S}'') - \sum[(\mathbf{S}'+\mathbf{S}'')\cdot\mathbf{I}_i]^2\}.$$

It may be shown that these inequalities correspond geometrically to the points on the hypercircle that are furthest from and closest to the origin.

VII

Variational Methods

VII–1. Principle of Virtual Work

In this chapter, a number of ways of reformulating the elasticity equations in terms of equivalent variational conditions will be discussed; of these, the simplest and most easily visualized physically is suggested by the classical principle of virtual work.

Consider any collection of material particles in equilibrium under the action of certain internal and external forces. Then the resultant of all internal and external forces acting on any one particle must be zero. If an imaginary small displacement is now given to each particle, then the work that would be done by this zero resultant force on each particle is zero, so that the total work that would be done by all forces on all particles is zero. The displacement is imaginary rather than real, for in any real displacement the various forces would certainly alter; to emphasize that each force is to remain unaltered in the calculation of work done, the displacement is conventionally described as *virtual*.

The total work done by all forces in a virtual displacement is then zero, by the requirement of equilibrium. This total work may be divided into two parts — that due to the external forces, and that due to the internal forces, this second part being merely the first-order variation in internal energy (we assume internal forces to be conservative). Thus the *principle of virtual work* may be stated as follows: If a collection of particles, possessing the property that the internal energy is a function of particle position only, is in equilibrium under the action of certain external forces, and if a virtual displacement is given to each particle, then the work done by the (unaltered) external forces acting through this displacement is equal to the (first-order) increase in internal energy.

This result applies directly and in a very general way to elastic bodies. For the present, only linear elasticity is being considered, and here the result may alternatively be obtained at once from the equilibrium equations. The strain energy ξ_e is given by the first of Eqs. (V–35), and the change (or variation) $\delta\xi_e$ corresponding to a virtual displacement δu_i (δu_i is a small displacement vector field) is:

$$\begin{aligned}\delta\xi_e &= \frac{E}{(1+\sigma)}\int\left[e_{ij}\,\delta e_{ij} + \frac{\sigma}{1-2\sigma}\,e_{ii}\,\delta e_{jj}\right]dV \\ &= \int \tau_{ij}\,\delta e_{ij}\,dV \\ &= \int \tau_{ij}\,\delta u_{i,j}\,dV \\ &= \int\Big[(\tau_{ij}\,\delta u_i)_{,j} - (\tau_{ij,j}\,\delta u_i)\Big]dV \\ &= \int T_i\,\delta u_i\,dS + \int \rho F_i\,\delta u_i\,dV, \end{aligned} \qquad \text{(VII–1)}$$

and this is indeed the principle of virtual work.

If, conversely, a displacement state u_i has been found such that Eq. (VII–1) is satisfied for all δu_i, then it will be shown that u_i is the solution of the elasticity problem. It is actually possible to prove a stronger theorem. Let the boundary condition be as in Sec. VI–5, and denote by S_u, S_τ those portions of the surface over which u_i, T_i are respectively prescribed. Then if a displacement state u_i has been found such that (1) it agrees with prescribed displacements $u_i^{(p)}$ on S_u, (2) Eq. (VII–1) is satisfied for all δu_i which vanish on S_u, then this u_i is the solution of the elasticity problem.

For let $T_i^{(p)}$ be the prescribed stress vector on S_τ, and let $F_i^{(p)}$ be the prescribed body force. Then we are told that

$$\delta\xi_e - \int T_i^{(p)}\,\delta u_i\,dS_\tau - \int \rho F_i^{(p)}\,\delta u_i\,dV = 0 \qquad \text{(VII–2)}$$

for all δu_i vanishing on S_u. If the stress tensor calculated by way of Hooke's law from $e_{ij} = \frac{1}{2}(u_{i,j} + u_{j,i})$ is denoted by τ_{ij}, then

$$\delta\xi_e = \int \tau_{ij}\,\delta u_{i,j}\,dV,$$

where the fact that $(\delta u_i)_{,j} = \delta(u_{i,j})$ (because of the linearity and consequent interchangeability of each operation) means that it is not necessary to distinguish between these two expressions. Continuing,

$$\delta\xi_e = \int \tau_{ij}n_j\,\delta u_i\,dS - \int \tau_{ij,j}\,\delta u_i\,dV,$$

where the first integral differs from zero only on S_τ, because of the restriction on δu_i. Equation (VII–2) therefore gives

$$\int (\tau_{ij}n_j - T_i^{(p)})\, \delta u_i\, dS - \int (\tau_{ij,j} + \rho F_i^{(p)})\, \delta u_i\, dV = 0 \quad \text{(VII–3)}$$

for any δu_i which vanishes on S_u. In particular, Eq. (VII–3) must hold for any δu_i which vanishes everywhere on S; for this case, the equation requires that

$$\int (\tau_{ij,j} + \rho F_i^{(p)})\, \delta u_i\, dV = 0.$$

If the integrand here were not identically zero, then a δu_i vanishing on S could be found so as to make the integral nonzero (cf. Sec. III–2). Consequently the equations of equilibrium are satisfied by the τ_{ij} corresponding to our u_i. The second integral in Eq. (VII–3) is therefore identically zero irrespective of what δu_i is, so that Eq. (VII–3) reduces to the statement that

$$\int (\tau_{ij}n_j - T_i^{(p)})\, \delta u_i\, dS = 0$$

for any δu_i vanishing on S_u. It again follows that the integrand must vanish, so that the boundary condition is met on S_τ; by hypothesis, the boundary condition is certainly met on S_u, so that u_i is now seen to satisfy all requirements (and therefore incidentally to be unique).

The process gone through here is typical of the calculus of variations, and will be used (without repeating the foregoing details) throughout this chapter. It will be noticed that the use of the divergence theorem so as to "replace" $\delta u_{i,j}$ by δu_i allowed us to avoid the inconvenience of having to vary δu_i and $\delta u_{i,j}$ simultaneously when trying to draw conclusions from Eq. (VII–3).

In the hypothesis of the converse theorem above, u_i was required to equal $u_i^{(p)}$ on S_u, but was free on S_τ. It was found, however, as a consequence of the fact that u_i was to satisfy the variational principle, that the stresses calculated from u_i by Hooke's law had to match $T_i^{(p)}$ on S_τ; such a boundary condition, which is not given initially but which results from a variational principle, is termed a *natural boundary condition.*

The fact that the differential equations of elasticity are equivalent to the principle of virtual work may be used in a number of ways — and similar remarks to those which follow may be made for any of the variational principles of this chapter.

First of all, the variational principle may often be used to simplify the derivation of the differential equations governing a certain problem. Consider, for example, the case of a thin plate. Here, the material fibers initially perpendicular to the middle surface are assumed for simplicity to remain so during the deformation (an assumption which is directly extrapolated from the engineering theory of beams). It is then desired to obtain the differential equation for the deflection of the middle surface. One way of doing this is to examine the equilibrium of a typical portion of the plate; however, it is simpler to express the elastic energy of the plate in terms of the deflection of the middle surface (using the assumption concerning perpendicularity of fibers) and then to apply the principle of virtual work to a small alteration in that deflection. This gives the differential equation directly.

Second, the variational principle may assist in the choice of appropriate boundary conditions. Again, the theory of thin plates may be used as an example. Physically, one would like to satisfy three boundary conditions for the plate (the details are not important here); however, a result of the simplifying assumptions is that the order of the differential equation for the middle-surface deflection turns out to be too low. When the variational principle is used, one finds that one of the two variational boundary conditions is the same as a previous one, but that the other one appears as a linear combination of two of the previous ones. The best choice of boundary conditions for the differential equation of the plate is therefore obtained by satisfying one of the original conditions and also a linear combination of the other two — and furthermore it is easy to see in this particular case that this is physically very reasonable.

Third, a variational principle is ideally suited for use with approximation methods, and this will be illustrated in the next section for the principle of virtual work.

Fourth, a variational principle may be used for deducing certain features of the solution for an elasticity problem without having to actually obtain the solution first. This fact will be illustrated in Sec. VII–4 in connection with Castigliano's theorem.

VII–2. Rayleigh-Ritz Methods

In the Rayleigh-Ritz method, a certain functional form is assumed for the quantity of interest (displacement vector in the general case, displacement of middle surface in the thin plate case of Sec. VII–1), and the principle of virtual work is then used to make the best choices for certain undetermined constants which have been incorporated into the functional form so as to allow some degree of flexibility. The intuitive feeling is that the fact that the variational theorem includes the entire body makes it the suitable tool for use in choosing these constants. Usually, the assumed function is in the form of a linear combination of a number of specific functions, the coefficients of the form being the constants in question. The variation δu is then obtained by altering slightly one coefficient at a time; since the strain energy is quadratic in these coefficients, the result of the variations will be a set of linear equations for the coefficients. The following remarks may be made:

(1) The situation is simplest if the boundary conditions on displacement are automatically satisfied by any choice of the undetermined constants (such as by using a series $\phi_0 + C_1\phi_1 + C_2\phi_2 + \cdots$, where ϕ_0 satisfies the boundary condition and where all other ϕ_i satisfy homogeneous boundary conditions). If this is not the case, it is necessary either to restrict the variations of the constants so as to be compatible with any boundary requirements, or (equivalently) to use the method of Lagrange multipliers to be discussed in Sec. VII–3.

(2) It may happen that a set of *complete* functions is known, that is, a set of functions such that any elasticity solution for the problem in question can be exactly expressed as a linear combination of those functions. If so, the choice of the coefficients by the variational principle will yield the exact solution, for no other choice of constants could satisfy the variational principle. This fact strengthens the intuitive feeling mentioned above to the effect that the variational principle is appropriate for the choice of the undetermined constants, for if only a few functions are used in an assumed functional form these functions may be thought of as being the first few of a complete series.

(3) It is clearly advantageous to choose the functions involved in an assumed linear combination to be mutually orthogonal in some sense. If, for example, no products of two different coefficients occur in the expression for the strain energy, then the values of the individual coefficients are obtained directly without the necessity of solving a set of linear equations.

(4) More generally, it is possible to use undetermined functions as well as undetermined constants in a particular functional form, and then to obtain the differential equation governing these undetermined functions by use of the variational principle.

(5) Unless the assumed functional form is such as to allow, by suitable choices of the undetermined quantities, reasonably close approximation

to the true solution, the Rayleigh-Ritz method may result in considerable error. Unless a complete set of functions is available as in (2) above, the only guide to an appropriate choice of functions is a combination of experience and intuition. In any event, however, it should be noticed that the principle of virtual work attempts to find a best fit for displacement rather than the various derivatives of displacement, so that one may generally expect less accuracy for stresses than for displacements. The opposite is true of the principle of virtual stress of Sec. VII–4.

It is not necessary to have available a variational principle of the elegance of the principle of virtual work in order to use the Rayleigh-Ritz method, for any set of differential equations may be squared and integrated over the region of interest in order to provide a variational principle. For example, consider the problem of finding a function $u(x)$ which satisfies the differential equation $L(u) = 0$ and which satisfies certain boundary conditions at two endpoints (a, b). Then alternatively, u is known to be that function satisfying the boundary conditions which also minimizes the expression

$$\int_a^b [L(u)]^2 \, dx. \tag{VII–4}$$

Writing

$$u = \phi_0 + \sum_1^n c_i\phi_i \tag{VII–5}$$

and choosing the c_i by the Rayleigh-Ritz method then gives an approximate solution which is in a least-squares sense the best fit to the differential equation (here ϕ_0 satisfies the given boundary conditions, the ϕ_i satisfy homogeneous boundary conditions). If Eq. (VII–5) is substituted into the expression (VII–4), this quantity becomes a function of the c_i, and may be minimized by differentiating with respect to each such c_i. The resulting set of linear equations for the c_i is

$$\int L(\phi_0 + \sum c_i\phi_i) L(\phi_j) \, dx = 0.$$

It has been suggested by Galerkin that it is computationally simpler, and with appropriate ϕ_i just about as accurate, to replace the $L(\phi_j)$ above by ϕ_j itself — that is, the *Galerkin procedure* is to orthogonalize the quantity $L(\phi_0 + \sum c_i\phi_i)$ with respect to the ϕ_i.

In the elastic problem presently being considered, when it is desired to find a displacement vector u_i satisfying certain boundary conditions and the equations of equilibrium, an equivalent of (VII–4) would be

$$\int \left(u_{i,jj} + \frac{1}{1-2\sigma} u_{j,ji} + \frac{1}{G} \rho F_i^{(p)}\right)\left(u_{i,ss} + \frac{1}{1-2\sigma} u_{s,si} + \frac{1}{G} \rho F_i^{(p)}\right) dV, \tag{VII–6}$$

and this expression is to be minimized for all u_i satisfying the boundary conditions on displacement and stress. The boundary restrictions may be removed by adding to the volume integral the surface integrals

$$\int (u_i^{(p)} - u_i)(u_i^{(p)} - u_i)\, dS_u + \int (T_i^{(p)} - \tau_{ij}n_j)(T_i^{(p)} - \tau_{is}n_s)\, dS_\tau, \tag{VII–7}$$

where τ_{ij} is calculated from u_i by Hooke's law. Variational principles obtained in this way have the same solution as the variational principle of virtual work and so must be transformable into it, and vice versa; however, the virtual-work form is usually simpler to use.

The addition of the expression (VII–7) to (VII–6) is an example of transforming prescribed boundary conditions into natural boundary conditions — something which can always be done and which is occasionally useful. It is also useful to note that in any variational principle the incorporation of an additional subsidiary condition which is known to be satisfied by the exact solution will not alter the solution of the principle; further, the addition of an integral whose first variation is known to vanish for the exact solution will not alter the solution of the variational principle. The first result has been used by Friedrichs [1] to transform a variational principle into a reciprocal one having the property that if the first principle relates to a maximum then the second relates to a minimum, and vice versa. The second has been used by Courant [2] in potential-theory problems to increase the rapidity of convergence of an approximating sequence by incorporating derivatives of higher order (which have an "ironing-out" effect) into the variational principle; because of the potential-theory nature of the equations of elasticity, a similar gain in rapidity of convergence may be expected here — at the expense, of course, of more labor. For quadratic functionals (such as occur in elasticity), the procedure of Friedrichs is equivalent to that of Sec. VII–6.

It need hardly be remarked that when approximate methods are to be

[1] K. O. Friedrichs, *Nachr. Ges. Wiss. Göttingen, Math.-physik. Kl.* (1929), p. 13.
[2] R. Courant, *Bull. Am. Math. Soc. 49* (1943), 1.

used it is worth while to consider — besides variational methods — some of the more conventional techniques such as collocation (the constants in an assumed solution are so chosen that the assumed solution fits perfectly at a certain number of points) or finite differences (cf. Sec. V–2).

VII–3. Lagrangian Multipliers

The basic idea of the method of Lagrangian multipliers may be illustrated by a simple example involving ordinary maximums or minimums. Let $f(x, y, z)$, $g(x, y, z)$ be two functions of the three variables x, y, z; it is desired to maximize (or, in any event, find the stationary points of) the function f, where x, y, and z are to be constrained by the condition $g = 0$. For example, it could be desired to maximize the volume $f = xyz$ of a parallelepiped whose sides are of length x, y, z, subject to the condition that the surface area is a constant, that is, $g = xy + yz + zx - C = 0$. It is possible to remove the restriction by solving the equation $g = 0$ for z in terms of x, y and then writing f in terms of the two free variables x, y; however, not only is this an unsymmetrical procedure (and so may be disadvantageous, in theoretical work at least) but it may not be analytically feasible. One can alternatively proceed as follows.

If f has a stationary value, then

$$f_x\,\delta x + f_y\,\delta y + f_z\,\delta z = 0 \tag{VII–8}$$

for all permissible alterations $\delta x, \delta y, \delta z$, where subscripts indicate partial derivatives. The alterations $\delta x, \delta y, \delta z$ are related by

$$g_x\,\delta x + g_y\,\delta y + g_z\,\delta z = 0. \tag{VII–9}$$

At the desired solution point, the six partial derivatives have certain numerical values, such that, for any $\delta x, \delta y, \delta z$ satisfying Eq. (VII–9), Eq. (VII–8) is also satisfied. Formally, δz may be expressed in terms of δx and δy by Eq. (VII–9) and the result substituted into Eq. (VII–8) to give

$$\left(f_x - \frac{f_z}{g_z}\,g_x\right)\delta x + \left(f_y - \frac{f_z}{g_z}\,g_y\right)\delta y = 0,$$

and since there are now no restrictions on $\delta x, \delta y$ it follows that

$$f_x - \left(\frac{f_z}{g_z}\right) g_x = 0,$$

$$f_y - \left(\frac{f_z}{g_z}\right) g_y = 0.$$

Also, identically,

$$f_z - \left(\frac{f_z}{g_z}\right) g_z = 0.$$

But the quantity in parentheses is simply a number (when evaluated at the desired solution point), and so may be treated as an unknown constant $-\lambda$. The last three equations then attain the symmetrical form

$$\begin{aligned} f_x + \lambda g_x &= 0, \\ f_y + \lambda g_y &= 0, \\ f_z + \lambda g_z &= 0. \end{aligned} \qquad \text{(VII–10)}$$

(Incidentally, had g_z vanished at the solution point so as to make the quantity f_z/g_z infinite, either g_y or g_x could have been used instead.) A useful way of looking at Eq. (VII–10) is that it is the same as the condition that the quantity $f + \lambda g$ be stationary, where λ is an unknown constant, and *where there are no restrictions on* δx, δy, δz. This last remark is the essence of the method of Lagrangian multipliers: the method is a formal device which removes restrictions on the variables, so that the usual differentiation (or variational) methods may be used. It is of course still necessary to adjoin the condition $g = 0$ to Eqs. (VII–10), but it is not necessary to use this condition in performing the differentiations of $f + \lambda g$.

In this parallelepiped example, we therefore know that there exists a certain constant λ such that

$$xyz + \lambda(xy + yz + zx - C)$$

is stationary, that is, such that its derivatives with respect to each of x, y, z vanish. Use of this fact together with $g = 0$ gives a set of four equations which may be solved for x, y, z; the optimal prism is found to be a cube.

The general case is handled similarly. If $f(x_1, x_2, \cdots, x_n)$ is to attain a stationary value, where the n variables are restricted by m conditions of the form

$$g_1(x_1, x_2, \cdots, x_n) = g_2(x_1, x_2, \cdots, x_n) = \cdots = g_m(x_1, x_2, \cdots, x_n) = 0,$$

then the various partial derivatives of the expression

$$f + \lambda_1 g_1 + \lambda_2 g_2 + \cdots + \lambda_m g_m$$

must vanish. This gives a set of n equations, which together with the m equations $g_i = 0$ is sufficient to determine the $m + n$ unknowns $x_1, x_2, \cdots, x_n, \lambda_1, \lambda_2, \cdots, \lambda_m$.

Consider now the variational problem. Let it be required to find those functions $y_1(x, y, z), y_2(x, y, z), \cdots, y_n(x, y, z)$ which make the quantity

$$\int f\left(x, y, z, y_1, y_2, \cdots, y_n, \frac{\partial y_1}{\partial x}, \cdots\right) dV \tag{VII–11}$$

stationary (such as a maximum or minimum) subject to the m conditions

$$\begin{aligned} g_1\left(x, y, z, y_1, y_2, \cdots, y_n, \frac{\partial y_1}{\partial x}, \cdots\right) &= 0, \\ g_m\left(x, y, z, y_1, y_2, \cdots, y_n, \frac{\partial y_1}{\partial x}, \cdots\right) &= 0. \end{aligned} \tag{VII–12}$$

Now any integral is the limit of a discrete sum, so that the expression (VII–11) may be approximated by the sum of a large number of quantities each of which depends upon the values of the y_i at various points (the derivatives being handled by finite differences). The values of the y_i at the points may be thought of as the independent variables, analogous to the x_i previously used. Following out this line of thought, it is easily seen that the Lagrangian multiplier method here consists in taking variations in the quantity

$$\int (f + \lambda_1 g_1 + \cdots + \lambda_m g_m)\, dV, \tag{VII–13}$$

where the λ_i are now functions of position. In the special case where one or more g_i are integral conditions (called isoperimetric conditions because of the historical association with the problem of maximizing the area of a figure of fixed perimeter), the corresponding λ_i are constants. For example, if the only subsidiary condition is $\int h_1\, dV = 0$, then the expression $\int (f + \lambda_1 h_1)\, dV$ is to be made stationary, where λ_1 is a constant.

It may be emphasized that, in these variational problems also, the gain in using Lagrangian multipliers is that the variations in the various functions are now completely free of any restrictions. Of course, once the variations have been carried out, the various subsidiary conditions must then be used in order to determine the λ_i.

In elasticity, the foregoing analysis is hardly necessary, for Lagrangian multipliers usually occur in a very natural and physically obvious manner. Again, the principle of virtual work will be used as an example. The principle of virtual work as given by Eq. (VII–1) holds for all δu_i, yet in using this principle (cf. Sec. VII–2), only such δu_i as vanished on S_u were considered (the converse theorem of Sec. VII–1 showing that this was sufficient). It may, however, be inconvenient to impose this restriction on the trial functions δu_i; alternatively, Eq. (VII–1) may be used directly and the δu_i may be considered to be completely free on the boundary. The stress vector T_i on S_u is not known, and so has the character of a Lagrangian multiplier function. After extracting whatever information is available by taking the free variation in Eq. (VII–1), one must then use the subsidiary condition $u_i = u_i^{(p)}$ on S_u. It is clear that the process is of the same nature as that leading to the expression (VII–13), the only difference being that the subsidiary condition here involves a surface rather than a volume integral. Note that the Lagrangian multiplier function has here a very definite physical interpretation, namely, the surface stress on S_u.

When the Rayleigh-Ritz method is being used, the introduction of Lagrangian multipliers presents no conceptual difficulty when a complete (in the sense of Sec. VII–2) series is being used. If, however, this is not the case, then the subsidiary condition cannot be met exactly and so must be approximated in some appropriate manner. Again, the advantage of using Lagrangian multipliers in a Rayleigh-Ritz problem is that the constants may be varied without regard to subsidiary conditions — that is, the variational equations are obtained by direct differentiation.

VII–4. Principle of Virtual Stress

The direct and in a sense reciprocal relation between stress and strain leads one to anticipate that variations in τ_{ij} rather than in u_i will also lead to useful results. Let ξ_τ denote the stress energy of Eq. (V–35); then, corresponding to $\delta\tau_{ij}$, a calculation of the same nature as that leading to Eq. (VII–1) now gives

$$\delta\xi_\tau = \int u_i\,\delta T_i\,dS + \int u_i\,\delta(\rho F_i)\,dV, \tag{VII–14}$$

where

$$\begin{aligned} \delta T_i &= (\delta\tau_{ij})n_j, \\ \delta(\rho F_i) &= -\,\delta\tau_{ij,j}, \\ \delta\tau_{ij} &= \delta\tau_{ji}. \end{aligned}$$

As in Sec. VII–1, a strong form of the converse of this *principle of virtual stress* may be proved. Suppose that a symmetric stress state τ_{ij} has been found such that $\tau_{ij}n_j = T_i^{(p)}$ on S_τ, $\tau_{ij,j} = -\rho F_i^{(p)}$ in V, and such that for all symmetrical $\delta\tau_{ij}$ satisfying $\delta\tau_{ij}n_j = 0$ on S_τ and $\delta\tau_{ij,j} = 0$ in V the equation

$$\delta\xi_\tau - \int u_i \delta T_i \, dS_u = 0 \tag{VII-15}$$

holds. Then this τ_{ij} state is the solution of the elasticity problem. For Eq. (VII–15) states that

$$\int e_{ij}\, \delta\tau_{ij}\, dV - \int u_i\, \delta\tau_{ij}n_j\, dS_u = 0$$

for all such $\delta\tau_{ij}$, where e_{ij} is defined from the given τ_{ij} state by means of Hooke's law. But, by the method of Sec. VI–5, e_{ij} can be written as $e_{ij}' + e_{ij}''$, where τ_{ij}' satisfies $\tau_{ij,j}' = 0$ in V and $\tau_{ij}'n_j = 0$ on S_τ, and $u_i'' = u_i^{(p)}$ on S_u. Then the last equation becomes

$$\int e_{ij}'\, \delta\tau_{ij}\, dV + \int u_{i,j}''\, \delta\tau_{ij}\, dV - \int u_i\, \delta\tau_{ij}n_j\, dS_u = 0,$$

that is,

$$\int e_{ij}'\, \delta\tau_{ij}\, dV = 0.$$

Choosing $\delta\tau_{ij} = \epsilon\tau_{ij}'$ (which by the definition of e_{ij}' is permissible) and using the positive definite nature of the strain energy shows that $e_{ij}' = 0$ so that $e_{ij} = e_{ij}''$, which is derivable from a displacement vector. In other words, the given τ_{ij} state satisfies the equations of compatibility, and so is the true (and unique) stress state.

A useful special case of Eq. (VII–14) is due to Castigliano. Suppose that a body, supported in some way, is subjected to a number of concentrated forces P_i and couples M_i, and let the strain energy ξ_τ be expressed in terms of these P_i and M_i. Consider first a $\delta\tau_{ij}$ which would correspond to a change only in P_1; then Eq. (VII–14) gives

$$\frac{\partial\xi_\tau}{\partial P_1}\, \delta P_1 = u_1\, \delta P_1,$$

where u_1 is the displacement component at the point of application of P_1 in the direction of P_1. Thus $u_1 = \partial\xi_\tau/\partial P_1$, and so forth, and similarly the rotations are given by $\theta_1 = \partial\xi_\tau/\partial M_1$, and so forth. If the loads are not concentrated, averaged results are obtained.

The principle of virtual stress may be used in the various ways discussed in Sec. VII–1 for the principle of virtual work. Rayleigh-Ritz methods as in Sec. VII–2 may also be used here. Lagrangian multipliers

(Sec. VII–3) now may be interpreted physically as unknown displacements.

If the Rayleigh-Ritz method is used with the principle of virtual work, then, since calculations are carried out in terms of displacements, the compatibility conditions are identically satisfied and the variational principle tries to satisfy the equilibrium equations as well as possible within the framework of the approximation being used. On the other hand, in the principle of virtual stress, the equilibrium conditions are identically satisfied and the principle tries to satisfy the compatibility conditions as well as possible. Reissner[3] has pointed out that it may on occasion be advantageous to treat the equations of compatibility and equilibrium on a more equal footing, and has suggested the following variational principle which achieves this result, at the expense, however, of having to consider variations in both stress and displacement. Consider the quantity

$$\int \tau_{ij} e_{ij} \, dV - \xi_\tau$$

and calculate its variation for arbitrary $\delta\tau_{ij}$, δe_{ij}. The result is

$$\int (\delta\tau_{ij} e_{ij} + \tau_{ij} \, \delta e_{ij}) \, dV - \int L(\tau_{ij}) \, \delta\tau_{ij} \, dV,$$

where $L(\tau_{ij})$ is the expression for strain e_{ij} in terms of stress. If the τ_{ij} and e_{ij} represent the true solution, then the first and third integrands cancel so that the expression becomes

$$\delta\left(\int \tau_{ij} e_{ij} \, dV - \xi_\tau\right) = \int T_i \, \delta u_i \, dS + \int \rho F_i \, \delta u_i \, dV. \qquad \text{(VII–16)}$$

Conversely, let τ_{ij} be a stress state which satisfies $\tau_{ij} n_j = T_i^{(p)}$ on S_τ, and let u_i be a displacement vector which satisfies $u_i = u_i^{(p)}$ on S_u. For all $\delta\tau_{ij}$ such that $\delta\tau_{ij} n_j = 0$ on S_τ, and for all δu_i such that $\delta u_i = 0$ on S_u, let

$$\delta\left(\int \tau_{ij} e_{ij} \, dV - \xi_\tau\right) = \int T_i^{(p)} \, \delta u_i \, dS + \int \rho F_i^{(p)} \, \delta u_i \, dV. \qquad \text{(VII–17)}$$

Then it follows easily that

$$e_{ij} = L(\tau_{ij}),$$
$$\tau_{ij,j} = -\rho F_i,$$

[3] E. Reissner, "On variational principles in elasticity," *Symposium on Calculus of Variations* (American Mathematical Society, Providence, Rhode Island, 1956).

so that compatibility and equilibrium both follow from the variational principle.

VII–5. Minimal Potential and Complementary Energy

The *potential energy* π is defined by

$$\pi = \xi_e - \int T_i u_i \, dS - \int \rho F_i u_i \, dV.$$

Here T_i and F_i are the true surface and body forces; u_i is any arbitrary vector field. Then the theorem of minimal potential energy states that the minimal value of π is attained for the true u_i. To prove this, let u_i be the true displacement, and $u_i + v_i$ any other. Let e_{ij}, f_{ij} be the strain tensors corresponding to u_i, v_i respectively. Then

$$\begin{aligned}\pi(u_i + v_i) - \pi(u_i) &= G \int \left[(e_{ij} + f_{ij})(e_{ij} + f_{ij}) + \frac{\sigma}{1 - 2\sigma} (e_{kk} + f_{kk})^2 \right] dV \\ &\quad - G \int \left(e_{ij} e_{ij} + \frac{\sigma}{1 - 2\sigma} e_{kk}{}^2 \right) dV \\ &\quad - \int T_i v_i \, dS - \int \rho F_i v_i \, dV \\ &= \int \tau_{ij} v_{i,j} \, dV + F - \int T_i v_i \, dS - \int \rho F_i v_i \, dV \\ &= F,\end{aligned}$$

where F is the positive definite strain energy corresponding to the displacement v_i. This proves the result.

The *complementary energy* π^* is defined by

$$\pi^* = \xi_\tau - \int T_i u_i \, dS - \int \rho F_i u_i \, dV,$$

where now u_i is the true displacement, and where τ_{ij} is any stress field T_i and F_i are the surface and body forces that would correspond to τ_{ij} Then, by the same type of proof, it is easily shown that the minimal value of π^* is attained for the true stress state.

Since the minimum in each case is attained for the true state, it follows that the first variation in each expression must vanish for variations from the true state of u_i, τ_{ij} respectively; this gives again the principles of virtual work and virtual stress. The converse theorems of Secs. VII–1 and VII–4 show, moreover, that not only do π, π^* have minima for the true states, but there are no other states for which these quantities are even stationary.

Because of the minimal theorems, any Rayleigh-Ritz or other approximations for the displacements or stresses must give too large a value — and so an upper bound — for π or π^* respectively. It may be noted that for the true solution $\pi = \pi^* = -\xi$, by Eq. (V–34).

VII–6 Function-Space Methods

The notation of Sec. VI–6 will be used. The inequalities there derived are of particular interest in connection with variational methods, for they can be used to calculate how far an approximate solution is from the true solution (distance being measured in terms of strain energy). It may be remarked that the results of Sec. VI–6 can also be derived by use of the methods of Sec. VII–5; to illustrate this, let us reconsider the inequality (VI–57):

$$\mathbf{S} \cdot \mathbf{S} \leq \mathbf{S}' \cdot \mathbf{S}' + \mathbf{S}'' \cdot \mathbf{S}''. \tag{VII–18}$$

The solution of an elastic problem may be written as the sum of two solutions. Let $\mathbf{S}_1$ be the solution of the problem $\tau_{ij,j} = -\rho F_i^{(p)}$, $\tau_{ij}n_j = T_i^{(p)}$ on S_T, $u_i = 0$ on S_u; let $\mathbf{S}_2$ be the solution of the problem $\tau_{ij,j} = 0$, $\tau_{ij}n_j = 0$ on S_T, $u_i = u_i^{(p)}$ on S_u. Then clearly $\mathbf{S} = \mathbf{S}_1 + \mathbf{S}_2$. Consider now any $\mathbf{S}'$ for the original problem; it is also an $\mathbf{S}'$ for the first subproblem, and, by the theorem of minimal complementary energy, $|\mathbf{S}_1|^2 \leq |\mathbf{S}'|^2$. Similarly, $|\mathbf{S}_2|^2 \leq |\mathbf{S}''|^2$. Now $\mathbf{S}_1 \cdot \mathbf{S}_2 = 0$, so that

$$\begin{aligned} |\mathbf{S}|^2 &= (\mathbf{S}_1 + \mathbf{S}_2) \cdot (\mathbf{S}_1 + \mathbf{S}_2) \\ &= |\mathbf{S}_1|^2 + |\mathbf{S}_2|^2 \\ &\leq |\mathbf{S}'|^2 + |\mathbf{S}''|^2, \end{aligned}$$

which is again Eq. (VII–18). Further, if $\mathbf{S}'$ coincides with $\mathbf{S}_1$ and $\mathbf{S}''$ with $\mathbf{S}_2$, then the minimum of the right-hand side is indeed attained — and it is interesting to note that the optimal value of the right-hand side is not $\mathbf{S}' = \mathbf{S}'' = \mathbf{S}$.

It has been pointed out by Courant [4] that the theorems of Sec. VII–5 are reciprocal in a way first discussed in general by Friedrichs. The principle of minimal complementary energy states that among all stress states $\mathbf{S}' + \mathbf{I}'$ (where $\mathbf{I}'$ is of the form $\sum k_p'\mathbf{I}_p'$ of Sec. VI–6) the quantity

$$\xi_T' - \int T_i' u_i \, dS_u$$

[4] R. Courant and D. Hilbert, *Methods of mathematical physics* (Interscience, New York, 1953), vol. 1, pp. 253, 268.

is a minimum for the true $\mathbf{I}'$. Here $\mathbf{S}'$ is some fixed state; let $\mathbf{S}''$ also be a fixed state. Then u_i in the integral over S_u may be replaced by u_i'', and also we can add the constant term

$$-\int T_i'u_i''\,dS_T - \int \rho F_i'u_i''\,dV + \xi_e'',$$

so that the quantity which is minimized by the true $\mathbf{I}'$ becomes

$$\xi_T' - \int T_i'u_i''\,dS - \int \rho F_i'u_i''\,dV + \xi_e''$$

or

$$\xi_T' - \int \tau_{ij}'e_{ij}'' + \xi_e'',$$

which may be written

$$\tfrac{1}{2}|\,\mathbf{S}' + \mathbf{I}'\,|^2 - (\mathbf{S}' + \mathbf{I}')\cdot\mathbf{S}'' + \tfrac{1}{2}|\,\mathbf{S}''\,|^2.$$

Consequently, the quantity

$$|\,\mathbf{S}' + \mathbf{I}' - \mathbf{S}''\,|^2 \tag{VII–19}$$

attains its minimum, d_1^2, for the true $\mathbf{I}'$. Similarly, the principle of minimal potential energy shows that the quantity

$$|\,\mathbf{S}'' + \mathbf{I}'' - \mathbf{S}'\,|^2, \tag{VII–20}$$

where I'' is of the form $\sum k_p''I_p''$, attains its minimum, d_2^2, for the true $\mathbf{I}''$. Since the solution is the same in each case, the following relation must hold between the true $\mathbf{I}'$ and $\mathbf{I}''$:

$$\mathbf{S}' + \mathbf{I}'_{\text{true}} = \mathbf{S}'' + \mathbf{I}''_{\text{true}}.$$

Note that the vectors $\mathbf{I}'$ and $\mathbf{I}''$ admitted in minimizing the quantities (VII–19) and (VII–20) belong to mutually orthogonal sets, for $\mathbf{I}'\cdot\mathbf{I}'' = 0$; nevertheless, $\mathbf{S}' + \mathbf{I}'_{\text{true}}$ belongs to the class $\mathbf{S}'' + \mathbf{I}''$, and vice versa. In this sense — in which the admissibility conditions of one are the solution conditions of the other — the problems of minimizing the quantities (VII–19) and (VII–20) are reciprocal. Note also that from Eq. (VI–48), replacing $\mathbf{S}$ by $\mathbf{S}' + \mathbf{I}'_{\text{true}}$, and so forth,

$$(\mathbf{S}' + \mathbf{I}'_{\text{true}} - \mathbf{S}'')\cdot(\mathbf{S}'' + \mathbf{I}''_{\text{true}} - \mathbf{S}') = 0,$$

from which

$$d_1^2 + d_2^2 = |\,\mathbf{S}' - \mathbf{S}''\,|^2. \tag{VII–21}$$

Thus if an upper bound is known for d_1^2, as obtained by use of any $\mathbf{I}'$ in the expression (VII–19), Eq. (VII–21) gives at once a lower bound for d_2^2.

This technique of obtaining upper and lower bounds on certain quantities is of general use in connection with various quadratic functionals of mathematical physics. The general analysis is formally identical; $\mathbf{S}'$ and $\mathbf{S}''$ are fixed vectors, and $\mathbf{I}'$, $\mathbf{I}''$ are chosen from mutually orthogonal sets.

Of more interest here is the problem of finding upper and lower bounds on stress or displacement at a point in terms of $\mathbf{S}'$ and $\mathbf{S}''$. Methods of doing this have been discussed by Diaz, Greenberg, Synge, Washizu, and others.

Let $u_i^{(q)}$ be a displacement vector satisfying zero body force except at an interior point ξ_i where there is a singularity of the type discussed in Sec. V–3. If quantities without superscripts refer to the solution of an elasticity problem whose boundary conditions are of the type of Sec. VI–5, then the reciprocal theorem gives

$$\int T_i^{(q)} u_i \, dS_0 - \int T_i u_i^{(q)} \, dS_0 = \int T_i u_i^{(q)} \, dS - \int T_i^{(q)} u_i \, dS + \int \rho F_i u_i^{(q)} \, dV, \tag{VII–22}$$

where dS_0 refers to a spherical surface surrounding the singularity. If this spherical surface is allowed to contract to a point, then first (for the $u_i^{(q)}$ contemplated here) the body force integral exists, and second, the left-hand side of the equation approaches a quantity P which may be calculated (by power-series expansions of the appropriate quantities about ξ_i) to be as follows:

(1) $u_i^{(q)} = (3 - 4\sigma) \dfrac{\delta_{i1}}{r} + \dfrac{x_1 x_i}{r^3}$; $P = 16(1 - \sigma)\pi G u_1(\xi_i)$;

(2) $u_i^{(q)} = e_{3it} \dfrac{x_t}{r^3}$; $P = 8\pi G \omega_{12}(\xi_i)$;

(3) $u_i^{(q)} = (2 - 4\sigma) \dfrac{x_1}{r^3} \delta_{i1} - (1 - 2\sigma) \dfrac{x_i}{r^3} + 3 \dfrac{x_1^2 x_i}{r^5}$; $P = 8(1 - \sigma)\pi \tau_{11}(\xi_i)$;

(4) $u_i^{(q)} = (1 - 2\sigma) \left(\dfrac{x_i}{r^3} - \dfrac{x_3 \delta_{i3}}{r^3} \right) + 3 \dfrac{x_1 x_2 x_i}{r^5}$; $P = 8(1 - \sigma)\pi \tau_{12}(\xi_i)$.

Here, r is the distance between ξ_i and x_i. These displacements will be recognized as due to a concentrated force, a concentrated couple, a double force without moment plus a center of expansive pressure, and a concentrated shear (that is, a pair of oppositely directed couples), respectively. (It is clear that these are the requisite choices of $u_i^{(q)}$ to yield quantities P of the kinds given.)

Define now $\mathbf{U}$ to be a stress state satisfying equilibrium with zero body force, and such that $T_i^{(U)} = -T_i^{(q)}$ on S_τ; define $\mathbf{V}$ to be a stress state derivable from a displacement vector which satisfies $u_i^{(V)} = -u_i^{(q)}$ on S_u. Then Eq. (VII–22) gives

$$P = \int T_i u_i^{(q)}\, dS_\tau - \int T_i^{(q)} u_i\, dS_u + \int \rho F_i u_i^{(q)}\, dV$$
$$+ \left(\int T_i u_i^{(V)}\, dS_\tau - \int T_i u_i^{(V)}\, dS \right) - \left(\int T_i^{(U)} u_i\, dS_u - \int T_i^{(U)} u_i\, dS \right),$$

where the two quantities in parentheses may be written

$$\int T_i u_i^{(V)}\, dS_\tau + \int \rho F_i u_i^{(V)}\, dV - \mathbf{S} \cdot \mathbf{V}$$

and

$$\int T_i^{(U)} u_i\, dS_u - \mathbf{S} \cdot \mathbf{U},$$

respectively. Consequently,

$$P = \int T_i (u_i^{(q)} + u_i^{(V)})\, dS_\tau - \int (T_i^{(q)} + T_i^{(U)}) u_i\, dS_u$$
$$+ \int \rho F_i (u_i^{(q)} + u_i^{(V)})\, dV + \mathbf{S} \cdot (\mathbf{U} - \mathbf{V}),$$

which has the form

$$P = A + \mathbf{S} \cdot (\mathbf{U} - \mathbf{V}). \qquad \text{(VII–23)}$$

Only $\mathbf{S}$ is unknown; however, we can write the last term in any of the ways

$$(\mathbf{S} - \mathbf{S}') \cdot (\mathbf{U} - \mathbf{V}) + \mathbf{S}' \cdot (\mathbf{U} - \mathbf{V}),$$
$$(\mathbf{S} - \mathbf{S}'') \cdot (\mathbf{U} - \mathbf{V}) + \mathbf{S}'' \cdot (\mathbf{U} - \mathbf{V}),$$
$$(\mathbf{S} - \tfrac{1}{2}[\mathbf{S}' + \mathbf{S}'']) \cdot (\mathbf{U} - \mathbf{V}) + \tfrac{1}{2}(\mathbf{S}' + \mathbf{S}'') \cdot (\mathbf{U} - \mathbf{V}),$$

and use the bounds of Sec. VI–6,

$$| \mathbf{S} - \mathbf{S}' | \le | \mathbf{S}' - \mathbf{S}'' |,$$
$$| \mathbf{S} - \mathbf{S}'' | \le | \mathbf{S}' - \mathbf{S}'' |,$$
$$| \mathbf{S} - \tfrac{1}{2}(\mathbf{S}' + \mathbf{S}'') | \le \tfrac{1}{2} | \mathbf{S}' - \mathbf{S}'' |,$$

to give bounds on the last term of Eq. (VII–23). But this is equivalent to obtaining upper and lower bounds on P, which is the desired result.

VIII

Thermoelasticity

VIII–1. Review of Thermodynamics

1. Heat and Temperature

Heat is a form of energy which is distinguished from electrical or mechanical energy by virtue of its molecularly disorganized nature. Experiment shows that it may be transformed into other forms of energy, and vice versa, in such a way as to maintain a fixed and transitive proportionality — that is, there is never any creation or destruction of energy. The temperature of a body is a measure of its heat content. Two bodies are defined to have the same temperature when they are in thermal equilibrium with each other in a state of thermal contact. One body is defined to have a higher temperature than another if heat will flow from the first to the second when the bodies are brought into thermal contact. Temperature is usually measured by use of a fluid, such as mercury or air, whose volume increases with increasing temperature.

2. Second Law of Thermodynamics

Experience, fortified by considerations of statistical mechanics, shows that it is not possible to construct a device which will — without other changes in the universe — have the net effect of transferring heat from a colder to a warmer body. This statement is adopted as a thermodynamic axiom, and is called the second law of thermodynamics. An equivalent statement is that it is not possible to transform heat completely into work, without other alteration in the universe. For if it were possible to do so, the resultant mechanical work could be transformed into heat (by friction, paddle wheels, electric resistance heating, and so on) at any desired temperature, and so the second law would be violated. It is important to note (as emphasized by Clausius in his original statement of the second law) that the second law does not say merely that heat will not flow from a colder to a warmer body; it says that it is not possible in any way by means of any intermediate device to trick it into so doing — without other changes in the universe.

3. *Reversibility*

In any heat engine, the process of obtaining mechanical work involves some "descent" of heat from a higher to a lower temperature. Thus, the burning of fuel provides heat at a high temperature which is used to create steam for a steam engine; after expanding in the cylinder and so providing mechanical work, the steam is discharged either to the atmosphere or to a condenser — and in either case heat is given up to low-temperature surroundings.

In attempting to decide what quantities controlled the efficiency of a heat engine working between two given temperatures, Carnot was led to the concept of a reversible engine, and was able to prove (by use of the caloric theory, in which heat was assumed to be an indestructible fluid) that no engine could have an efficiency greater than that of a reversible one. Clausius obtained the same result by use of the second law.

A reversible process is defined to be one which can be reversed by only infinitesimal changes in the external universe. An example is the slow and careful isothermal expansion of a gas in contact with a thermal reservoir; if the process is reversed, exactly the same mechanical work has to be put back into the gas as was originally obtained, and exactly the same heat is given back to the reservoir as was previously obtained from it. On the other hand, the frictional heating of two blocks sliding relative to one another is not reversible, for the second law prohibits the transformation of heat into work without other net effect.

Consider now two heat engines, A and R, each extracting heat at a temperature T_1 and each discharging to a lower temperature T_2. Let R be a reversible engine. For convenience, let a complete cycle of each engine (in which whatever working substance is used returns to its final state) correspond to the production of W units of mechanical work. Let the heat taken from the high-temperature reservoir be $Q_1^{(A)}$, $Q_1^{(R)}$ respectively. Then Carnot's theorem states that

$$\frac{W}{Q_1^{(R)}} \geq \frac{W}{Q_1^{(A)}},$$

and the proof consists simply in noting that, if the opposite were the case, then allowing engine A to drive R backward would have the net result of transferring heat from T_2 to T_1 — a violation of the second law.

4. Thermodynamic Temperature

Kelvin noticed that the efficiency of a reversible engine must, because of the foregoing result, depend only on the temperatures T_1, T_2 and not on the working substance used, and hence that this fact provided a method of defining temperature which was independent of the properties of any particular substance. In order to make this thermodynamic temperature coincide as closely as possible with previous measures of temperature, Kelvin defined the ratio of two such *absolute* temperatures to be given by

$$\frac{T_2}{T_1} = 1 - \eta = \frac{Q_2}{Q_1}, \tag{VIII–1}$$

where Q_1, Q_2 are the quantities of heat absorbed at T_1 and given out at T_2 by a reversible engine of efficiency η working between these two temperatures; this definition still leaves undetermined a possible constant multiplier, which is chosen so that the difference in temperature between the ice and steam points of water is 100° (centigrade absolute) or 180° (Fahrenheit absolute). Thus, if a reversible engine is operated between any two temperatures, Eq. (VIII–1) gives the ratio of the corresponding absolute temperatures. In this way (noting that Eq. (VIII–1) is consistent for various combinations of temperatures), all temperatures may be obtained within some constant multiplier, which is then chosen as above. It is found that the absolute temperature of the ice point is 273° (absolute centigrade, or °K) or 491° (absolute Fahrenheit, or °R). Since all temperatures are positive by Eq. (VIII–1), it follows that there is a lower limit of absolute temperature, namely, 0°A or 0°R. A consistent absolute temperature scale could have been devised (related exponentially to the present one) such that there was no lower limit of temperature; however, such a scale would not have coincided (at least approximately) with ordinary mercury or gas measurements of temperature.

5. Entropy

Consider any reversible process which traverses a closed cycle, so that a particular thermodynamic system is brought back to its initial state. During this process, heat will have been absorbed by the system at various temperatures and emitted at other temperatures, and mechanical work will also have been exchanged with the rest of the universe. Let an amount of heat δQ_i (positive if absorbed, negative if emitted) be ex-

changed between system and external universe at a temperature T_i.
Then it will be proved that

$$\sum \frac{\delta Q_i}{T_i} = \int \frac{dQ}{T} = 0 \qquad \text{(VIII–2}$$

for the complete cycle.

To prove this, let the system be carried through exactly the same close process; however, whenever heat δQ_i is to be absorbed, let it be done by means of an auxiliary reversible engine which absorbs heat $\delta Q_i T_1/T_i$ a some fixed high temperature T_1 and delivers it (in a single cycle) at th requisite T_i; similarly, whenever heat is to be emitted, let it be give to an auxiliary reversible engine which (during a single cycle) transform part of it into work and which emits the remainder $|\delta Q_i| T_2/T_i$ at low temperature T_2. The combination of all these reversible processe is still a reversible process which now operates between T_1 and T_2 further, a closed cycle has been performed, and so the theorem of (4 above gives

$$\frac{\sum \delta Q_i T_1/T_i}{\sum |\delta Q_i| T_2/T_i} = \frac{T_1}{T_2},$$

so that (taking account of sign)

$$\sum \frac{\delta Q_i}{T_i} = 0.$$

The result (VIII–2) may be used to define a state function S calle *entropy*. Consider any two states of a thermodynamic system. Then th difference in entropies of these two states is defined as

$$S_2 - S_1 = \int \frac{dQ}{T}$$

calculated along some reversible path joining the two states (no longer closed circuit, in general). That the same entropy difference would b found for some other reversible path between the same two states follow from the fact that the combination of the two paths — one forward an one backward — forms a reversible complete cycle to which Eq. (VIII–2 may be applied. Thus, if the entropy of some reference state is given a arbitrary value, the entropy of any other state is defined in a way whic depends only on that state and not on the (reversible) path used to ge there — that is, entropy is a state function.

From the viewpoint of statistical mechanics, entropy can be related to the degree of disorganization of the motions of the molecules constituting the system.

As an example of the calculation of entropy, consider a simple homogeneous substance (such as a gas or a piece of metal) of volume V and temperature T, subjected to a pressure P. Let the only mechanical work exchange with the surroundings be by means of pressure-volume changes. If now a slow and careful (reversible) process is carried out whereby the volume increases by dV and the internal energy U by dU, the amount of heat added must be

$$dQ = dU + P\,dV,$$

so that

$$T\,dS = dU + P\,dV, \qquad \text{(VIII-3)}$$

which is one of the most important relations in all thermodynamics. Although this relation was derived on the basis of a reversible differential process, it expresses a relation between differentials of state functions and so holds for all processes, reversible or not.

VIII–2. Thermodynamics of a Simple Substance

Defining a simple substance as in the last paragraph of Sec. VIII–1, denote the pressure, volume, temperature, entropy, and internal energy by P, V, T, S, U respectively; let any two of these be enough to determine the state. Then there must, for example, exist a function f such that

$$f(P, V, T) = 0$$

and this implies that

$$\left(\frac{\partial f}{\partial P}\right)_{V,T}\left(\frac{\partial P}{\partial V}\right)_T + \left(\frac{\partial f}{\partial V}\right)_{P,T} = 0,$$

and so forth, from which the useful relation

$$\left(\frac{\partial P}{\partial V}\right)_T\left(\frac{\partial V}{\partial T}\right)_P\left(\frac{\partial T}{\partial P}\right)_V = -1 \qquad \text{(VIII-4)}$$

is obtained. Similar results hold for any other combination of three variables. Note also that

$$\left(\frac{\partial P}{\partial V}\right)_T = \frac{1}{\left(\dfrac{\partial V}{\partial P}\right)_T}$$

and so forth, and that

$$\left(\frac{\partial P}{\partial V}\right)_T = \frac{(\partial P/\partial S)_T}{(\partial V/\partial S)_T},$$

and so forth.

There are also a number of specific relations which hold as a result of Eq. (VIII–3). This equation implies that, treating U as a function of S and V,

$$\left(\frac{\partial U}{\partial S}\right)_V = T \quad \text{and} \quad \left(\frac{\partial U}{\partial V}\right)_S = -P.$$

But because $\partial^2 U/\partial V \partial S$ is independent of the order of partial differentiation, it then follows that

$$\left(\frac{\partial T}{\partial V}\right)_S = -\left(\frac{\partial P}{\partial S}\right)_V. \tag{VIII–5}$$

Also, the three quantities H (enthalpy), F (free energy), and G (free enthalpy) may be defined by

$$\begin{aligned} H &= U + PV, \\ F &= U - TS, \\ G &= U + PV - TS. \end{aligned}$$

Then use of Eq. (VIII–3) shows that

$$\begin{aligned} dH &= T\,dS + V\,dP, \\ dF &= -S\,dT - P\,dV, \\ dG &= -S\,dT + V\,dP, \end{aligned}$$

from which, in exactly the same way as for Eq. (VIII–5), the remaining three of *Maxwell's equations* may be written down:

$$\left(\frac{\partial T}{\partial V}\right)_S = -\left(\frac{\partial P}{\partial S}\right)_V, \tag{VIII–5}$$

$$\left(\frac{\partial T}{\partial P}\right)_S = \left(\frac{\partial V}{\partial S}\right)_P, \tag{VIII–6}$$

$$\left(\frac{\partial S}{\partial V}\right)_T = \left(\frac{\partial P}{\partial T}\right)_V, \tag{VIII–7}$$

$$\left(\frac{\partial S}{\partial P}\right)_T = -\left(\frac{\partial V}{\partial T}\right)_P. \tag{VIII–8}$$

Certain partial derivatives are more easily measured experimentally han others, and these are conventionally designated by special symbols. The first such is the thermal coefficient of volume expansion, defined by

$$\alpha = \lim_{\delta V \to 0} \left(\frac{\delta V / V}{\delta T} \right)_P = \frac{1}{V} \left(\frac{\partial V}{\partial T} \right)_P .$$

The second is the specific heat at constant pressure C_P, defined by

$$C_P = \lim_{\delta T \to 0} \left(\frac{\delta Q}{\delta T} \right)_P ,$$

nd, since $\delta Q = T\,\delta S$, this can be written

$$C_P = T \left(\frac{\partial S}{\partial T} \right)_P ,$$

which form has the advantage of involving only state functions. Similarly, he specific heat at constant volume is

$$C_V = T \left(\frac{\partial S}{\partial T} \right)_V .$$

The isothermal bulk modulus is defined by

$$\kappa_T = -V \left(\frac{\partial P}{\partial V} \right)_T ,$$

nd the adiabatic bulk modulus by

$$\kappa_S = -V \left(\frac{\partial P}{\partial V} \right)_S ,$$

where the fact that S is held constant in the last definition implies that, n a reversible process during which the pressure is increased, no heat is o be allowed to pass into or out of the specimen. The negative signs are ncorporated into the last two definitions in order to make the coefficients ositive.

It is now possible to derive a number of physically interesting relaions:

$$(a) \quad \left(\frac{\partial P}{\partial T} \right)_V = - \left(\frac{\partial V}{\partial T} \right)_P \left(\frac{\partial P}{\partial V} \right)_T \quad \text{by Eq. (VIII–4)}$$

$$= \alpha \kappa_T ; \qquad \text{(VIII–9)}$$

$$(b)\qquad \frac{C_P}{C_V} = \frac{(\partial S/\partial T)_P}{(\partial S/\partial T)_V} = \frac{(\partial P/\partial T)_S(\partial S/\partial P)_T}{(\partial V/\partial T)_S(\partial S/\partial V)_T}$$

$$= \left(\frac{\partial P}{\partial V}\right)_S \left(\frac{\partial V}{\partial P}\right)_T$$

$$= \frac{-V(\partial P/\partial V)_S}{-V(\partial P/\partial V)_T} = \frac{\kappa_S}{\kappa_T}; \qquad \text{(VIII–10)}$$

$$(c)\quad C_P - C_V = T\left[\left(\frac{\partial S}{\partial T}\right)_P - \left(\frac{\partial S}{\partial T}\right)_V\right]$$

$$= T\left\{\left(\frac{\partial S}{\partial T}\right)_P - \left[\left(\frac{\partial S}{\partial T}\right)_P + \left(\frac{\partial S}{\partial P}\right)_T \left(\frac{\partial P}{\partial T}\right)_V\right]\right\}$$

$$= T\left(\frac{\partial V}{\partial T}\right)_P \left(\frac{\partial P}{\partial T}\right)_V$$

$$= TV\alpha^2\kappa_T; \qquad \text{(VIII–11)}$$

$$(d)\qquad \kappa_S - \kappa_T = (C_P - C_V)\frac{\kappa_T}{C_V}$$

$$= \frac{TV\alpha^2\kappa_T{}^2}{C_V}; \qquad \text{(VIII–12)}$$

$$(e)\qquad dU = T\,dS - P\,dV$$

$$= T\left[\left(\frac{\partial S}{\partial T}\right)_V dT + \left(\frac{\partial S}{\partial V}\right)_T dV\right] - P\,dV$$

$$= C_V\,dT + (T\alpha\kappa_T - P)\,dV. \qquad \text{(VIII–13)}$$

For metals, the quantities V, α, C_P, C_V, κ_T, κ_S are largely independent of pressure. Empirically, the quantity $V\alpha^2\kappa_T/C_P{}^2$ is almost independent of temperature, over reasonable ranges, say from room temperature to 1000°F. Some typical room-temperature numerical values for steel are

$$\alpha = 1.95 \times 10^{-5}/°R,$$
$$\kappa_T = 22 \times 10^6 \text{ lb/in.}^2,$$
$$C_P = 93.4 \text{ ft lb/lb } °R,$$
$$V = 3.52 \text{ in.}^3/\text{lb}.$$

Then Eq. (VIII–11) gives

$$C_P - C_V = 1.29 \text{ ft lb/lb } °R,$$

and Eq. (VIII–12) gives

$$\kappa_S - \kappa_T = 0.31 \times 10^6 \text{ lb/in.}^2.$$

These various coefficients are temperature dependent; at 1000°F, the difference between κ_S and κ_T is about 10 percent.

It is interesting to note that the apparent heat generated in the adiabatic compression of a piece of steel may be much more than the mechanical work done, the difference arising from a change in internal energy. For example, consider the effect of an adiabatic increase in pressure of 100 atm applied to 1 lb of steel initially at 68°F. The temperature rise is

$$dT = \left(\frac{\partial T}{\partial P}\right)_S dP = -\left(\frac{\partial S}{\partial P}\right)_T \left(\frac{\partial T}{\partial S}\right)_P dP$$

$$= \left(\frac{\partial V}{\partial T}\right)_P \left(\frac{\partial T}{\partial S}\right)_P dP = \frac{TV\alpha}{C_P}\, dP, \qquad \text{(VIII–14)}$$

which in this case equals

$$dT = \frac{(459 + 68)(3.52/12^3)(1.95 \times 10^{-5})}{(93.4)}(100)(14.7)(12^2)$$

$$= 0.0475°\text{F},$$

where dP could be replaced by the total change in pressure because of the fact that the coefficient of dP was essentially independent of pressure. To heat the steel by the same amount would require a heat input of

$$C_P\, dT = (93.4)(0.0475) = 4.43 \text{ ft lb},$$

whereas the amount of mechanical work actually done was

$$-\int P\, dV = \frac{V}{\kappa_S} \int P\, dP = 0.014 \text{ ft lb}.$$

By use of Eq. (VIII–14), it may now be shown that the difference between the pressures required for equal adiabatic and isothermal compression, as given by Eq. (VIII–12), corresponds to that amount which is required to suppress the thermal expansion caused by adiabatic heating. To prove this, apply dP adiabatically to give

$$dV = -\frac{V}{\kappa_S}\, dP,$$

$$dT = \frac{TV\alpha}{C_P}\, dP.$$

The heating dT is responsible for a part $\alpha V\, dT$ of dV, so that the volume changed produced by isothermal compression would have been

$$-\frac{V}{\kappa_S}\, dP - \frac{TV^2\alpha^2}{C_P}\, dP.$$

Consequently,

$$\kappa_T = \frac{dP}{\left(-\frac{1}{V}\right)\left(-\frac{V}{\kappa_S} - \frac{TV^2\alpha^2}{C_P}\right) dP},$$

and this agrees exactly with Eq. (VIII–12), thus verifying this interpretation of the difference between κ_S and κ_T.

VIII–3. Adiabatic and Isothermal Elastic Coefficients

It was shown in Sec. VIII–2 that the difference between κ_S and κ_T could be interpreted in terms of the temperature change associated with the former. A more direct way in which to obtain this result is to calculate directly the volume changes dV_S, dV_T resulting from the same dP, for the conditions of constant entropy and constant temperature, respectively:

$$dV_S = \left(\frac{\partial V}{\partial P}\right)_S dP,$$

$$dT_S = \left(\frac{\partial T}{\partial P}\right)_S dP,$$

$$dV_T = \left(\frac{\partial V}{\partial P}\right)_T dP.$$

If the foregoing interpretation is to hold, we must then have

$$dV_S = dV_T + \left(\frac{\partial V}{\partial T}\right)_P dT_S$$

that is,

$$\left(\frac{\partial V}{\partial P}\right)_S = \left(\frac{\partial V}{\partial P}\right)_T + \left(\frac{\partial V}{\partial T}\right)_P \left(\frac{\partial T}{\partial P}\right)_S,$$

and since this is an obvious identity (think of V as a function of P and T for which the partial derivative on the left-hand side of the equation is to be calculated), the result is indeed correct.

The same sort of result holds in general. The state of any elastic body is completely known if the stresses τ_{ij} and temperature T are known. If first a certain stress state is applied adiabatically, then a certain adiabatic deformation state will result; also, the temperature will increase by dT (which may of course be positive or negative). The body can be brought to precisely the same final state by first applying the stress state isothermally (thus attaining a certain isothermal deformation state) and then altering the temperature by the value of dT above — which results in thermal expansion. Thus the following statement may be made: If stresses are applied adiabatically to a body, then the deformation may be calculated in either of two equivalent ways: (1) by use of the elasticity equations relating stress and strain, in which the adiabatic elastic constants are used; (2) by use of the same equations involving now the isothermal elastic constants, together with a thermal-expansion term corresponding to the adiabatic temperature increase.

In Sec. VIII–2 the relation between the adiabatic and isothermal bulk moduli was discussed; since any two of the elastic constants determine all others, it is only necessary to consider one more such constant, and we will choose E. Consider a bar of length L and cross-sectional area A, subject to an axial tensile load F. Then Eq. (VIII–3) must be replaced by

$$T\,dS = dU - F\,dL \tag{VIII–15}$$

and Maxwell's equations become

$$\begin{aligned}
\left(\frac{\partial T}{\partial L}\right)_S &= \left(\frac{\partial F}{\partial S}\right)_L, \\
\left(\frac{\partial T}{\partial F}\right)_S &= -\left(\frac{\partial L}{\partial S}\right)_F, \\
\left(\frac{\partial S}{\partial L}\right)_T &= -\left(\frac{\partial F}{\partial T}\right)_L, \\
\left(\frac{\partial S}{\partial F}\right)_T &= \left(\frac{\partial L}{\partial T}\right)_F.
\end{aligned} \tag{VIII–16}$$

Define

$$\begin{aligned}
\alpha_L &= \frac{1}{L}\left(\frac{\partial L}{\partial T}\right)_F, \\
C_F &= T\left(\frac{\partial S}{\partial T}\right)_F, \\
C_L &= T\left(\frac{\partial S}{\partial T}\right)_L, \\
E_T &= \frac{L}{A}\left(\frac{\partial F}{\partial L}\right)_T, \\
E_S &= \frac{L}{A}\left(\frac{\partial F}{\partial L}\right)_S,
\end{aligned}$$

where these quantities may be expected to depend on temperature but not on stress. There is no practical error in assuming $\alpha_L = \frac{1}{3}\alpha$. As in Eqs. (VIII–9) to (VIII–12), we obtain

$$\begin{aligned} &(a) & \left(\frac{\partial F}{\partial T}\right)_L &= -\alpha_L A E_T, \\ &(b) & \frac{C_F}{C_L} &= \frac{E_S}{E_T}, \\ &(c) & C_F - C_L &= T L \alpha_L^2 A E_T, \\ &(d) & E_S - E_T &= \frac{T A L \alpha_L^2 E_T^2}{C_L}. \end{aligned} \qquad \text{(VIII–17)}$$

The temperature change due to an adiabatic dF is

$$dT = -\frac{T L \alpha_L}{C_F}\, dF, \qquad \text{(VIII–18)}$$

and, using this result, the same interpretation for $E_S - E_T$ may be made here as was made in the last section for $\kappa_S - \kappa_T$.

Equation (VIII–18) gives

$$dT = -\frac{T V \alpha}{3 C_F} \frac{dF}{A};$$

if the same value of dF/A is applied to each face of a prism of volume V — corresponding to a pressure $(-\,dP)$ — then by superposition the total change in temperature is

$$dT = \frac{T V \alpha}{C_F}\, dP$$

and comparison with Eq. (VIII–14) indicates that $C_F = C_P$. Using this result for the value of C_F, Eq. (VIII–18) predicts that a steel bar loaded in tension up to the elastic limit will experience a temperature drop of 0.32°F.

Of course, the values of C_F and C_P do depend slightly on the stress; the result $C_F = C_P$ is exactly true only at zero stress. To examine the stress dependence, consider for example C_F. Writing S as a function of (F, T) and denoting the value of C_F corresponding to a tensile force F by $C_F(F)$,

$$\begin{aligned} C_F(F) &= T\left(\frac{\partial S}{\partial T}\right)_F \\ &= T\left[\left(\frac{\partial S}{\partial T}\right)_{F=0} + \left(\frac{\partial^2 S}{\partial T \partial F}\right)_{F=0} F + \cdots\right], \end{aligned}$$

where a power-series expansion is being used. But, by Eq. (VIII–16),

$$TF \frac{\partial}{\partial T}\left(\frac{\partial S}{\partial F}\right) = TF \frac{\partial}{\partial T}\left\{\left(\frac{\partial L}{\partial T}\right)_F\right\}_F$$

$$= TF \frac{\partial}{\partial T}(L\alpha_L),$$

so that

$$C_F(F) = C_F(0) + TF \frac{\partial}{\partial T}(L\alpha_L) + \cdots.$$

Using the structural-steel value of $\alpha_L = (0.63 \times 10^{-5}) + (2.2t \times 10^{-9})$, where t is the temperature in °F, the last equation may be used to show that the difference between the values of C_F at zero stress and at the elastic-limit stress is about one part in a thousand.

It is not true, however, that C_L and C_V are equal, even at zero stress (this is reasonable, as they involve different constraining conditions). For Eqs. (VIII–11) and (VIII–17c) may be combined to give

$$C_F - C_L = \frac{1}{9}\frac{E_T}{\kappa_T}(C_P - C_V),$$

and thus, although $C_F = C_P$, it does not follow that $C_L = C_V$.

By use of equations of the kind derived in this section, all possible thermoelastic coefficients may be expressed in terms of E_T, κ_T, α, and C_P (the last being sufficiently accurately the specific heat at any constant stress state). In particular,

$$\frac{G_S}{G_T} = \frac{3E_S\kappa_S/(9\kappa_S - E_S)}{3E_T\kappa_T/(9\kappa_T - E_T)}$$

$$= \frac{C_P}{C_L}\frac{C_P}{C_V}\frac{9\kappa_T - E_T}{9(C_P/C_V)\kappa_T - (C_P/C_L)E_T}$$

and using the immediately previous equation for the ratio (E_T/κ_T), we obtain $G_S/G_T = 1$. The physically plausible fact that $G_S = G_T$ can also be proved directly. Consider a stress state in which only the two principal stresses τ_1 and τ_2 are different from zero. Then in an adiabatic change of stress

$$dT = \left(\frac{\partial T}{\partial \tau_1}\right)_S d\tau_1 + \left(\frac{\partial T}{\partial \tau_2}\right)_S d\tau_2.$$

But these two coefficients (for zero stress) are identical, by isotropy; also, a pure shear state requires $d\tau_1 = -\,d\tau_2$, so that it then follows that

$dT = 0$. Consequently there is no temperature change in adiabatic shear, so that an adiabatic shear is also an isothermal shear (at least, in linear elasticity).

The foregoing discussion of the elastic constants rested on separate thermodynamic analyses of the cases of pure compression and pure tension. It is of course also possible to treat the general situation directly. Let a parallelepiped be chosen whose sides are of lengths L_1, L_2, L_3 and which is lined up with the principal axes of stress and strain. Then if the tensile forces are F_1, F_2, F_3, the equivalent of Eq. (VIII–3) is

$$T\,dS = dU - F_1\,dL_1 - F_2\,dL_2 - F_3\,dL_3 \qquad \text{(VIII–19)}$$

and from this a large number of Maxwell-type equations can be written:

$$\left(\frac{\partial T}{\partial L_1}\right)_{S,L_2,L_3} = \left(\frac{\partial F_1}{\partial S}\right)_{L_1,L_2,L_3},$$

$$\left(\frac{\partial T}{\partial F_1}\right)_{S,L_2,L_3} = -\left(\frac{\partial L_1}{\partial S}\right)_{F_1,L_2,L_3},$$

and so on. From these and Sec. VIII–1, a number of relations between various special thermal and elastic constants may be obtained. However, the main results of interest in elasticity have already been derived in this section, so that Eq. (VIII–19) will not be pursued further except to note that if the parallelepiped is not lined up with the principal axes then Eq. (VIII–19) becomes

$$T\,dS = dU - V\tau_{ij}\,de_{ij}.$$

VIII–4. Mixed Thermoelastic Equations

Consider a general isotropic elastic body in which the stress and temperature distribution may be changing with time. In time dt, let the stress at a point alter by $d\tau_{ij}$. Consider a small parallelepiped of volume V lined up with the principal axes at this point; Eq. (VIII–18) together with superposition shows that an apparent amount of heat

$$C\,dT = -\frac{TV\alpha}{3}\,d\tau_{ii}$$

is generated by the stress increase. Here C is, with sufficient exactness, any of the specific heats; if c is the specific heat per unit mass, then $C = \rho Vc$. This amount of heat may be thought of as having been given to the material by an outside source. Now the discussion of Secs. VIII–2 and VIII–3 shows that the alteration in strain corresponding to $d\tau_{ij}$ is the same as if the process were first carried out isothermally and an amount of heat $C\,dT$ then added; however, before writing down this result it is necessary to note that heat may also have been conducted into or out of the prism during the time dT. If K is the conduction coefficient, this amount of heat is $KVT_{,ii}\,dt$. Consequently the net value of dT is

$$dT = -\frac{TV\alpha}{3C}\,d\tau_{ii} + \frac{KVT_{,ii}}{C}\,dt \tag{VIII–20}$$

and the increase in strain is

$$de_{ij} = \frac{1+\sigma_T}{E_T}\,d\tau_{ij} - \frac{\sigma_T}{E_T}\,\delta_{ij}\,d\tau_{kk} + \delta_{ij}\,\frac{\alpha}{3}\,dT. \tag{VIII–21}$$

Setting $i = j$ in Eq. (VIII–21) and using the result to eliminate $d\tau_{kk}$ in Eq. (VIII–20) gives

$$\frac{\partial T}{\partial t}\left(1 - \frac{T\alpha^2}{3c\rho}\,\frac{E_T}{1-2\sigma_T}\right) = \frac{K}{\rho c}\,T_{,kk} - \frac{T\alpha}{3c\rho}\left(\frac{E_T}{1-2\sigma_T}\right)\frac{\partial u_{k,k}}{\partial t}, \tag{VIII–22}$$

whereas integrating Eq. (VIII–21) and using the equations of equilibrium gives

$$u_{i,jj} + \frac{1}{1-2\sigma}\,u_{j,ji} = -\frac{\rho}{G_T}\,F_i + \frac{\rho}{G_T}\,\frac{\partial^2 u_i}{\partial t^2} + \frac{\alpha(2+2\sigma_T)}{3(1-2\sigma_T)}\,T_{,i}. \tag{VIII–23}$$

In Eq. (VIII–22), the second term on the left-hand side is usually less than 0.02 and so may usually be neglected; the last term on the right-hand side represents the temperature increase due to straining, and by Secs. VIII–2 and VIII–3 this may usually also be omitted. The equation then reduces to the usual one for the distribution of heat in a body, and so the problem becomes "unmixed" since Eq. (VIII–22) may be solved first and then Eq. (VIII–23). In a vibrating elastic body, the presence of the mixed terms in the equations above will result in damping — a fact which is physically obvious, since temperature increases resulting from local compression will give rise to heat conduction away from that region.

VIII–5. Conventional Thermoelastic Problem

The previous results indicate that it is usually not necessary to consider the mixed term in the equation governing temperature distribution. If the mixed term is omitted, the temperature distribution can be found by conventional methods and may be considered known. For a stationary isotropic medium (that is, one that is isotropic for both elastic and thermal effects), let T be the temperature level above some base, where T

will in general vary from point to point. Then the strain will be composed of two parts, one due to the stress τ_{ij} and one due to thermal expansion associated with T. If α is the *volume* coefficient of thermal expansion,

$$e_{ij} = \frac{1+\sigma}{E}\,\tau_{ij} - \frac{\sigma}{E}\,\delta_{ij}\tau_{kk} + \frac{\alpha}{3}\,\delta_{ij}T, \qquad \text{(VIII–24)}$$

from which

$$\tau_{ij} = \frac{E}{1+\sigma}\left(e_{ij} + \frac{\sigma}{1-2\sigma}\,\delta_{ij}e_{kk}\right) - \frac{\alpha ET\delta_{ij}}{3(1-2\sigma)}. \qquad \text{(VIII–25)}$$

Define an artificial stress τ_{ij}' by

$$\tau_{ij}' = \frac{E}{1+\sigma}\left(e_{ij} + \frac{\sigma}{1-2\sigma}\,\delta_{ij}e_{kk}\right), \qquad \text{(VIII–26)}$$

so that τ_{ij}' is related to the strain tensor e_{ij} by Hooke's law, and τ_{ij}' would therefore correspond to the true stress τ_{ij} if there were no thermal effects. The equation of equilibrium reads

$$\tau_{ij,j} + \rho F_i = 0,$$

which becomes

$$\tau_{ij,j}' + \rho\left[F_i - \frac{E\alpha T_{,i}}{3\rho(1-2\sigma)}\right] = 0. \qquad \text{(VIII–27)}$$

The quantity in brackets will be considered to be an artificial body force F_i'. The surface-stress vector that would match τ_{ij}' is

$$T_i' = \tau_{ij}'n_j = \left[\tau_{ij} + \frac{\alpha ET\delta_{ij}}{3(1-2\sigma)}\right] n_j$$

$$= T_i + \frac{\alpha ET}{3(1-2\sigma)}\, n_i \qquad \text{(VIII–28)}$$

Thus (τ_{ij}', e_{ij}) is the solution of an ordinary elasticity problem for which the body force is F_i' and the surface stress is T_i' (wherever prescribed). Note incidentally that

$$\int T_i'\,dS + \int \rho F_i'\,dV = 0,$$

$$\int e_{ijk}x_jT_k'\,dS + \int e_{ijk}x_j\rho F_k'\,dV = 0,$$

so that over-all equilibrium is satisfied. Having found τ_{ij}', τ_{ij} is given by Eq. (VIII–25).

It may be noted that τ_{ij}' satisfies the Beltrami-Michell compatibility equations (with F_i' substituted for F_i); τ_{ij} on the other hand does not. Note also that no stresses would arise from the (unconfined) uniform or linear heating of an elastic body.

Because of the equivalence of thermal stress to a fictitious body force, all previous elastic results (reciprocal theorem, concentrated force solutions, removal of body force, general solutions, variational methods, and so on) can be applied here also; the effect of temperature can be exhibited specifically if desired.

IX

Time-Dependent Problems

IX–1. Elastic Waves

Elastic waves may be discussed in terms of either stresses or displacements, but for general considerations a choice of the latter is simpler because fewer equations are involved. For small displacements, Navier's equations become

$$u_{i,jj} + \frac{1}{1 - 2\sigma} u_{j,ji} = \frac{\rho}{G}\frac{\partial^2 u_i}{\partial t^2} - \frac{\rho F_i}{G}, \tag{IX–1}$$

where it has been assumed that d^2u_i/dt^2 can be replaced by $\partial^2 u_i/\partial t^2$ — an assumption which may be verified a posteriori where necessary. For the most part, we will take $F_i = 0$. The special case of a plane wave helps to elucidate the structure of these equations. Consider an initial disturbance confined to some region of a large elastic body. The disturbance will gradually spread out, and a long way from the initially disturbed region will have the local character of an approximately plane wave moving perpendicularly to its wave front. A perfect plane wave is an idealization of this situation. Choose a coordinate system such that the x_1-axis is in the direction of advance; then nothing can depend on x_2 or x_3, so that Eqs. (IX–1) become

$$\begin{aligned} \left(\frac{2 - 2\sigma}{1 - 2\sigma}\right) u_{1,11} &= \frac{\rho}{G}\frac{\partial^2 u_1}{\partial t^2}, \\ u_{2,11} &= \frac{\rho}{G}\frac{\partial^2 u_2}{\partial t^2}, \\ u_{3,11} &= \frac{\rho}{G}\frac{\partial^2 u_3}{\partial t^2}. \end{aligned} \tag{IX–2}$$

Each of these equations has the form of the wave equation

$$c^2 \frac{\partial^2 \phi}{\partial x^2} = \frac{\partial^2 \phi}{\partial t^2}, \tag{IX–3}$$

whose general solution is easily found. By use of the coordinate transformation

$$\xi = x + ct,$$
$$\eta = x - ct,$$

Eq. (IX–3) becomes

$$\frac{\partial^2 \phi}{\partial \xi \partial \eta} = 0,$$

so that

$$\begin{aligned} \phi &= f(\xi) + g(\eta) \\ &= f(x + ct) + g(x - ct), \end{aligned} \qquad \text{(IX–4)}$$

which is a combination of two waves of fixed shape. One wave moves to the right with speed c, the other to the left with speed c. In the case of the outward-moving waves represented by Eqs. (IX–2), an observer would see one wave in which u_1 alone altered, moving with velocity

$$c_1 = \left[\left(\frac{2 - 2\sigma}{1 - 2\sigma}\right)\frac{G}{\rho}\right]^{\frac{1}{2}},$$

and another wave in which u_2 and u_3 alone altered, moving with the smaller velocity

$$c_2 = \left(\frac{G}{\rho}\right)^{\frac{1}{2}}.$$

It will be noticed that the longitudinal wave is irrotational, the transverse solenoidal. If the solution for u_1 is written $g(x - c_1 t)$, then the particle velocity is

$$\frac{du_1}{dt} \simeq \frac{\partial u_1}{\partial t} = -c_1 g',$$

and so is not the same as the wave speed c_1. The condition that the approximation made in the last equation be valid is that $g' \ll 1$.

In the general case, there are two combinations of the u_i which satisfy equations analogous to Eqs. (IX–2). Taking the divergence and curl of Eq. (IX–1) gives, respectively,

$$c_1^2 u_{i,ijj} = \frac{\partial^2}{\partial t^2} u_{j,j}, \qquad \text{(IX–5)}$$

$$c_2^2 \omega_{ij,kk} = \frac{\partial^2}{\partial t^2} \omega_{ij}, \qquad \text{(IX–6)}$$

so that each of the four quantities $u_{i,i}$, ω_{ij} satisfies a wave equation of the form

$$c^2\phi_{,jj} = \frac{\partial^2 \phi}{\partial t^2}. \tag{IX–7}$$

The wave nature of this equation is illustrated by the fact that one solution of the spherically symmetric case is easily checked to be

$$\phi = \frac{1}{r} f(r - ct), \tag{IX–8}$$

which represents an expanding wave of constant speed c, of decreasing amplitude, and of slightly altering shape.

Other solutions of Eq. (IX–7) are more complicated; however, they have in common with Eq. (IX–8) the fact that disturbances are propagated with speed c. This may be seen either by a little experimentation with the difference-equation approximation to Eq. (IX–7), or by use of the results of Sec. IX–2. Equation (IX–7) is one which often arises in mechanics (as in vibration of a string, membrane, sphere) and physical considerations indicate that a solution is uniquely determined by (*a*) specifying initial values of ϕ and $\partial\phi/\partial t$ throughout the spatial region, and (*b*) prescribing some boundary conditions (for example, the value of ϕ at all t) for the boundary of the spatial region. A formal solution of Eq. (IX–7) satisfying such initial and boundary conditions will be obtained in Sec. IX–3.

In an elasticity problem, the initial values of u_i and $\partial u_i/\partial t$ (and therefore of $u_{i,i}$, ω_{ij}, $\partial u_{i,i}/\partial t$, $\partial \omega_{ij}/\partial t$) are presumably known. If the boundary conditions are now such that Eqs. (IX–5) and (IX–6) can be solved so as to give the values of $u_{i,i}$ and ω_{ij} at all t, the problem then arises of determining u_i itself from these known values of its divergence and curl. Let u_i' be any vector function (considered known) which has these values of divergence and curl. Then

$$u_i = u_i' + \psi_{,i} \tag{IX–9}$$

where ψ is harmonic; the value of ψ must now be determined. Substitution of Eq. (IX–9) into (IX–1) gives

$$u_{i,jj}' + \frac{1}{1 - 2\sigma} u_{j,ji}' = \frac{\rho}{G}\left(\frac{\partial^2 u_i'}{\partial t^2} + \frac{\partial^2 \psi_{,i}}{\partial t^2}\right). \tag{IX–10}$$

But this means that the expression

$$u_{i,jj}' - \frac{\rho}{G}\left(\frac{\partial^2 u_i'}{\partial t^2}\right) + \frac{1}{1 - 2\sigma} u_{j,ji}'$$

is the gradient of some function ω, and, since u_i' is known, ω may be considered known. Then Eq. (IX–10) may be integrated to give

$$\frac{\rho}{g}\left(\frac{\partial^2 \psi}{\partial t^2}\right) = \omega, \tag{IX–11}$$

where an arbitrary function of time arising from the integration has been dropped because it would not show up in Eq. (IX–9) in any event. The right-hand side of Eq. (IX–11) being known, ψ may now be found by direct integration with respect to time, the arbitrary functions arising from this integration being determined from the known initial values of u_i.

Consider now a point P outside the region of an initially localized disturbance in an elastic body. The propagation of the curl and divergence of u_i are governed by Eqs. (IX–5) and (IX–6) and so until the first of these signals — traveling with velocity c_1 — reaches P, the curl and divergence of u_i at P remain zero. Then u_i is at most the gradient of a harmonic function, and an argument of the kind used in the previous paragraph shows that u_i is zero. Consequently no disturbance can be felt at P before the first of the waves (IX–5), (IX–6) arrives.

It may be noted that if the motion is solenoidal, so that $u_{i,i} = 0$, then Eq. (IX–1) gives

$$c_2^2 u_{i,jj} = \frac{\partial^2 u_i}{\partial t^2},$$

and if it is irrotational, so that $\omega_{ij} = 0$, whence $u_i = \Omega_{,i}$, then Eq. (IX–1) gives

$$c_1^2 u_{i,jj} = \frac{\partial^2 u_i}{\partial t^2}.$$

Thus a disturbance may propagate at the speed c_2 rather than c_1; moreover, it follows from the previous paragraph that any disturbance surface must propagate normal to itself into the undisturbed body at a local velocity which is either c_1 or c_2, so that no other velocity is possible.

It will now be shown that any solution of Eq. (IX–1) can be written in a particularly simple form. From Chapter I,

$$u_i = \Phi_{,i} + e_{ijk} A_{k,j}$$

where Φ and A_k are time-dependent functions. One such set of Φ and A_i having been chosen, a more general representation is

$$u_i = (\Phi + g)_{,i} + e_{ijk}(A_k + G_k + S_{,k})_{,j},$$

where g is harmonic and

$$g_{,i} + e_{ijk}G_{k,j} = 0.$$

For a given harmonic g, the divergence of $g_{,i}$ vanishes, so that such a G_k may always be found. Use of Eq. (IX–5) now shows that if g is chosen such that

$$\frac{\partial^2 g}{\partial t^2} = c_1^2 \Phi_{,jj} - \frac{\partial^2}{\partial t^2}\Phi$$

(the right-hand side is a specific harmonic function; that there exists a harmonic g satisfying this equation can be seen by writing the formal solution for g and taking its Laplacian) then the quantity

$$\phi = \Phi + g$$

satisfies

$$c_1^2 \phi_{,ii} = \frac{\partial^2}{\partial t^2}\phi.$$

Define $B_i = A_i + G_i$; then substituting

$$u_i = \phi_{,i} + e_{ijk}(B_k + S_{,k})_{,j}$$

into Eq. (IX–1) gives

$$\text{curl}\left(c_2^2 B_{i,jj} - \frac{\partial^2}{\partial t^2} B_i\right) = 0,$$

so that the quantity in parentheses is the gradient of some function H. Choosing S to satisfy

$$c_2^2 S_{,jj} - \frac{\partial^2}{\partial t^2} S = -H$$

(and physically we know that the wave equation has a solution for any reasonable forcing function) then means that the quantity

$$C_i = B_i + S_{,i}$$

satisfies

$$c_2^2 C_{i,jj} = \frac{\partial^2}{\partial t^2} C_i.$$

Consequently, any solution of the elastic-wave equation can be written as

$$u_i = \phi_{,i} + e_{ijk}C_{k,j},$$

vhere

$$c_1{}^2\phi_{,jj} = \frac{\partial^2}{\partial t^2}\phi,$$

$$c_2{}^2C_{i,jj} = \frac{\partial^2}{\partial t^2}C_i.$$

Finally, a discussion of energy flux in elastic waves is of interest. It vas shown in Sec. V–7 that the rate at which surface and body forces vere doing work on any portion of an elastic body in motion equaled the ate of accumulation of strain and kinetic energy of that portion. If there re no body forces, the surface-traction term may be thought of as repre- enting a flux of energy; consequently we define an energy-flux vector by

$$E_i = -\tau_{ji}\frac{\partial u_j}{\partial t},$$

o that the rate at which energy is flowing out of a part of the body is

$$\int E_j n_j \, dS.$$

or the plane waves of Eqs. (IX–2), the rate at which energy is being ransferred across a unit area normal to the x_1-axis is $\rho c_1{}^3(f')^2$ and $\rho c_2{}^3(f')^2$ espectively; for waves moving from left to right, the energy flows from eft to right. It is also easily checked that for either of these waves the nergy is at any instant half potential and half kinetic.

X–2. Characteristic Surfaces

In investigating properties of the wave equation (IX–7), several nethods of attack lead to a consideration of the same set of geometric urves or surfaces which seem to be closely associated with general prop- rties of solutions, and are therefore termed *characteristics*. One very nat- ral way of encountering these surfaces is to ask if there are any (x_i, t) urfaces across which solutions may have discontinuities of one sort or nother, say, for example, a discontinuity in $\partial^3\phi/\partial x_1{}^3$. Any disturbance r signal must after all involve some sort of discontinuity, so that the ime-space path of a disturbance would then have to lie on such a surface.

Consider first the one-dimensional equation (IX–3). The region of in- erest will be a certain portion of the (x, t) plane (we choose the x-axis orizontal, the t-axis vertical). Are there any curves

$$\xi(x, t) = \text{constant}$$

across which some partial derivative could be discontinuous? Choose a new curvilinear coordinate net

$$\xi = \xi(x, t),$$
$$\eta = \eta(x, t),$$

where it is proposed to make the curves of constant ξ those across which discontinuities can occur. Then Eq. (IX–3) becomes (using subscripts to indicate partial differentiation)

$$(c^2\xi_x^2 - \xi_t^2)\phi_{\xi\xi} = (2\xi_t\eta_t - 2c^2\xi_x\eta_x)\phi_{\xi\eta} + (\eta_t^2 - c^2\eta_x^2)\phi_{\eta\eta} + (\xi_{tt} - c^2\xi_{xx})\phi_\xi + (\eta_{tt} - c^2\eta_{xx})\phi_\eta. \quad \text{(IX–12)}$$

Suppose now that $\phi_{\xi\xi\xi}$ were to be (possibly) discontinuous across ξ = constant, with all lower ξ-derivatives continuous, and with all η-derivatives continuous. Then differentiate Eq. (IX–12) with respect to ξ to give

$$(c^2\xi_x^2 - \xi_t^2)\phi_{\xi\xi\xi} = \text{other terms}, \quad \text{(IX–13)}$$

where the other terms are continuous across ξ = constant. Writing Eq. (IX–13) for each of two infinitesimally separated points adjoining the curve on opposite sides to one another, and subtracting, gives

$$(c^2\xi_x^2 - \xi_t^2)[\phi_{\xi\xi\xi}] = 0$$

where the bracketed expression is the difference of $\phi_{\xi\xi\xi}$ at the two points. If there is to be a discontinuity, this difference cannot vanish, so that

$$c^2\xi_x^2 - \xi_t^2 = 0,$$

that is,

$$c\xi_x = \pm\, \xi_t. \quad \text{(IX–14)}$$

Consequently the contour lines of ξ in the (x, t) plane are lines of constant $(x - ct)$ or of constant $(x + ct)$. Thus the time-space path of any signal must be along one or the other of these straight lines, which are the characteristic curves of the present problem. We may choose

$$\xi = x + ct,$$
$$\eta = x - ct,$$

and then the lines of constant ξ or constant η are the characteristic lines. (If c were not constant, but a function of (x, t), then the characteristic curves would have slopes $dx/dt = \pm\, c$ and so would no longer be straight lines.)

Suppose now that values of ϕ and ϕ_t are known for a portion of the x-axis, that is, initial values of ϕ and ϕ_t are prescribed for this portion o the space region. Consider a point whose coordinates are (x, t); form a triangle by drawing through (x, t) the two characteristics which link it to the x-axis. Also, drop a perpendicular from (x, t) onto the x-axis; points along this perpendicular represent the same spatial position for various values of time. Then the value of ϕ at (x, t) can depend only on the signals that have been received at this x-position for all previous time. Such signals can only follow characteristics, and the only characteristics which the point (x) has seen are those which cut the perpendicular, that is, those which lie within the triangle. Consequently, the value of ϕ at (x, t) can depend only on the data prescribed for that part of the x-axis which forms the base of the triangle. Similarly, if the data are known *only* for this part of the x-axis, ϕ can be determined only for those points (x, t) lying within the triangle. Another way of saying this is that the value of (ϕ, ϕ_t) at some point Q on the x-axis can influence only those points in the (x, t) plane which lie between the two characteristics through Q. For the present straight-line characteristic case, the solution can be exhibited analytically. From Eq. (IX–4),

$$\phi = f(\xi) + g(\eta),$$

so that

$$\phi(x, 0) = f(x) + g(x),$$
$$\phi_t(x, 0) = cf'(x) - cg'(x),$$

from which

$$\begin{aligned} f(\xi) &= \frac{1}{2}\phi(\xi, 0) + \frac{1}{2c}\int_0^\xi \phi_t(\lambda, 0)\, d\lambda + K, \\ g(\eta) &= \frac{1}{2}\phi(\eta, 0) - \frac{1}{2c}\int_0^\eta \phi_t(\lambda, 0)\, d\lambda - K, \end{aligned} \tag{IX–15}$$

and

$$\phi(x, t) = \frac{1}{2}\Big[\phi(x + ct, 0) + \phi(x - ct, 0)\Big] + \frac{1}{2c}\int_{x-ct}^{x+ct} \phi_t(\lambda, 0)\, d\lambda. \tag{IX–16}$$

In the special case $\phi_t(x, 0) = 0$, an initial ϕ distribution thus splits into two equal halves which then travel with fixed shape and speed in opposite directions.

The situation is a little more complicated if the spatial region is bounded. For example, let $\phi(0, t)$ be prescribed to vanish for all t (for instance, the left-hand end of a string is fixed) and consider certain prescribed values of ϕ and ϕ_t on the positive half of the x-axis. For that portion of the positive (x, t) quadrant to the right of the line $x = ct$, $f(\xi)$

and $g(\eta)$ are still given by Eq. (IX–15) and $\phi(x, t)$ by Eq. (IX–16). In the region to the left of the line $x = ct$, the lines of constant ξ run back into the first region, where $f(\xi)$ is known; consequently $f(\xi)$ in the new region is still given by the first of Eqs. (IX–15). However, $g(\eta)$ is as yet not known, for the lines of constant η do not intersect the known region. On the t-axis, we have

$$f(\xi) + g(\eta) = 0,$$

where $\xi = ct$ and $\eta = -ct$; using the known value of $f(\xi)$ gives

$$\begin{aligned} g(\eta) &= -f(-\eta) \\ &= -\frac{1}{2}\phi(-\eta, 0) - \frac{1}{2c}\int_0^{-\eta} \phi_t(\lambda, 0)\, d\lambda - K, \end{aligned} \qquad \text{(IX–17)}$$

so that

$$\phi(x, t) = \frac{1}{2}\Big[\phi(x + ct, 0) - \phi(ct - x, 0)\Big] + \frac{1}{2c}\int_{ct-x}^{x+ct} \phi_t(\lambda, 0)\, d\lambda, \qquad \text{(IX–18)}$$

and this is the required value of $\phi(x, t)$ in the new region. The reflection character of this wave is particularly clear for the case $\phi_t(x, 0) = 0$. It is useful to note that the boundary condition at $x = 0$ could be replaced by an appropriate choice of ϕ, ϕ_t for the negative part of the x-axis; if the x-region were bounded at two ends, then either this image technique or a step-by-step filling up of the (x, t) space with characteristics could be used.

Since the solution was completely determined by use of the known value of ϕ on the t-axis, an attempt to prescribe both ϕ and its normal derivative ϕ_x on the t-axis would not have been successful. A general statement is that if characteristics cut the boundary more than once, then Cauchy conditions (prescribe ϕ and $\partial\phi/\partial n$) can be given only for a part of the boundary; Dirichlet (prescribe ϕ) or Neumann (prescribe $\partial\phi/\partial n$) conditions — or a single combination of the form $\alpha\phi + \beta\partial\phi/\partial n$ — are sufficient for the rest of the boundary. Note also that had we prescribed each of ϕ, $\partial\phi/\partial n$ along the characteristic $x = ct$, and, given no other information, the solution in the foregoing problem could not have been found for any region of the (x, t) plane.

If more than one space dimension is involved, the general situation is quite similar. Consider the two-dimensional problem

$$c^2(\phi_{xx} + \phi_{yy}) = \frac{\partial^2\phi}{\partial t^2},$$

and choose a new curvilinear coordinate system (ξ, η, ζ) which is to have the property that there can be discontinuities across a surface of constant ξ. Then, as in Eq. (IX–13),

$$[c^2(\xi_x^2 + \xi_y^2) - \xi_t^2]\phi_{\xi\xi\xi} = \text{other terms},$$

from which

$$c^2(\xi_x^2 + \xi_y^2) - \xi_t^2 = 0 \tag{IX–19}$$

if ξ = constant is to be a characteristic surface. The vector (ξ_x, ξ_y, ξ_t) is perpendicular to the surface ξ = constant; Eq. (IX–19) implies that this vector lies in the surface of a circular cone, and hence the surfaces of constant ξ must be circular cones whose generators make angles $\tan^{-1} c$ with their axes, these axes being normal to the (x, y) plane. Analogously to the one-dimensional case, the following remarks are a consequence of the fact that signals can be propagated only along characteristic surfaces:

(1) The value of ϕ at a point $P(x, y, t)$ depends only on the values of ϕ and ϕ_t on that part of the (x, y) plane intercepted by the characteristic cone through P.

(2) The values of ϕ and ϕ_t at a point Q on the (x, y) plane can affect only those points lying within the characteristic cone through Q.

Again, one would normally expect to have Cauchy conditions prescribed over a portion of the (x, y) plane, and Dirichlet or Neumann conditions around the boundary of that portion. However, since only one family of characteristics exists, it is not possible to build up a solution by their use in the same way as in the case of one space dimension. It may be noted that the characteristic cones cut the (x, t) plane in two families of straight lines, which are the characteristics for the one-dimensional case.

Finally, the results for three space dimensions are similar; the equation satisfied by the four-dimensional vector normal to a characteristic surface is

$$c^2(\xi_x^2 + \xi_y^2 + \xi_z^2) = \xi_t^2,$$

so that the surface itself is a hypercone whose projections onto hyperplanes of constant t are spheres.

IX–3. General Solutions of the Wave Equation

It was pointed out by Beltrami that the wave equation (IX–7) attains a rather interesting form if the substitution

$$\psi(x_i, t) = \phi\left(x_i, t - \frac{r}{c}\right) \tag{IX–20}$$

is made, where r is the distance from the origin to the point (x_i). That is, a function ψ is defined which has the same value at any point and time as ϕ had a certain time earlier. Equation (IX–7) becomes

$$\frac{1}{r}\psi_{,ii} + \frac{2}{c}\left(\frac{x_i}{r^2}\frac{\partial\psi}{\partial t}\right)_{,i} = 0. \qquad \text{(IX–21)}$$

In this form, only a single time-derivative occurs; also, the equation involves the divergence of a certain quantity and so makes the application of the divergence theorem useful. Note first that

$$\frac{1}{r}\psi_{,ii} = \left[\frac{1}{r}\psi_{,i} - \left(\frac{1}{r}\right)_{,i}\psi\right]_{,i} + \psi\left(\frac{1}{r}\right)_{,ii}$$

Using this substitution, integrate Eq. (IX–21) throughout a volume surrounding the origin. The quantity $(1/r)_{,ii}$ vanishes except at the origin, where it acts like a delta-function operator (see Sec. VI–1); the final result is

$$4\pi\psi(0, t) = \int\left[\frac{1}{r}\frac{\partial\psi}{\partial n} - \psi\frac{\partial}{\partial n}\left(\frac{1}{r}\right) + \frac{2}{cr}\frac{\partial r}{\partial n}\frac{\partial\psi}{\partial t}\right] dS.$$

The left-hand side is directly $4\pi\phi(0, t)$, since $r = 0$ at the origin. The right-hand side may be expressed in terms of ϕ by use of Eq. (IX–20). Defining the "retarded potential" of any quantity Ω to be

$$[\Omega] = \Omega\left(x_i, t - \frac{r}{c}\right),$$

and noting that

$$\frac{\partial\psi}{\partial n} = \left[\frac{\partial\phi}{\partial n}\right] - \frac{1}{c}\frac{\partial r}{\partial n}\left[\frac{\partial\phi}{\partial t}\right],$$

the result becomes

$$4\pi\phi(0, t) = \int\left\{\frac{1}{rc}\frac{\partial r}{\partial n}\left[\frac{\partial\phi}{\partial t}\right] + \frac{1}{r}\left[\frac{\partial\phi}{\partial n}\right] - \frac{\partial}{\partial n}\left(\frac{1}{r}\right)[\phi]\right\} dS, \qquad \text{(IX–22)}$$

which is Kirchhoff's formula. Thus the value of ϕ at any point P and at any time t may be obtained in terms of the values of ϕ, $\partial\phi/\partial t$, and $\partial\phi/\partial n$ on any surface S surrounding P; these values must be evaluated at a previous time, the time difference r/c being that required for a signal to reach P from the part of the surface in question. Equation (IX–22) is a form of Huygen's principle familiar from optics, wherein a certain geometric method is used to generate a subsequent wave front from a preceding one. A discussion of the relation is given by Coulson.[1] If the

[1] C. A. Coulson, *Waves* (Oliver and Boyd, London, 1952), pp. 142 ff.

surface S surrounding P is a sphere, then Eq. (IX–22) is easily transformed into a result obtained earlier by Poisson. Let the values of ϕ and ϕ_t at $t = 0$ be denoted by $p(x_i)$, $q(x_i)$ respectively. Then the value of ϕ at time t and at a point P is given in terms of integrals over a sphere of radius $r = ct$ by

$$4\pi r^2 \phi(P, t) = t \int q \, dS + \int \frac{\partial}{\partial r} (pr) \, dS. \qquad \text{(IX–23)}$$

Equation (IX–22) holds also for the case in which the origin is outside S, provided that the "exterior normal" now points into the interior of S, and that the various integrals vanish for a surface at infinity. This result may be used to show that a signal pulse is not followed by a wave train. Consider a single spherical pulse which has been generated in some manner, and which at a certain time is nonzero only for those radial distances from the origin which lie between certain limits. Draw a surface S between the pulse and the origin. Then, using Eq. (IX–22) for any point P outside S, it follows that there is no disturbance at P for some period following passage of the pulse. On the other hand, the two-dimensional equivalent of Eq. (IX–22) involves area rather than contour integrals, so that two-dimensional signals are followed by wave trains.

In wave-equation problems, it is possible to use Green's functions in a manner analogous to that of Chapter VI. In Sec. VI–1, it may be seen that the Green's function for Laplace's equation satisfies

$$\nabla^2 G(x;\xi) = -4\pi \, \delta(x_1 - \xi_1) \, \delta(x_2 - \xi_2) \, \delta(x_3 - \xi_3),$$

where the delta function $\delta(x)$ is defined to vanish except at $x = 0$ where it gives a unit impulse, so that for any $f(x)$

$$\int_{-\alpha}^{+\alpha} f(x) \, \delta(x) \, dx = f(0).$$

The function $\delta(x)$ may be thought of as the limit as $\epsilon \to 0$ of the well behaved function $\epsilon/\pi(x^2 + \epsilon^2)$. Similarly, the Green's function for the wave equation is defined so as to correspond to a concentrated source:

$$\nabla^2 G(x,t;\xi,\tau) - \frac{1}{c^2} \frac{\partial^2}{\partial t^2} G(x,t;\xi,\tau)$$

$$= -4\pi \, \delta(x_1 - \xi_1) \, \delta(x_2 - \xi_2) \, \delta(x_3 - \xi_3) \, \delta(t - \tau). \qquad \text{(IX–24)}$$

Here the point (ξ_i) and time τ represent the source point and time, and are considered fixed in Eq. (IX–24). To complete the specification of G, we arbitrarily set G and $\partial G/\partial t$ equal to zero for $t < \tau$. Then by Fourier-transform methods (as described in texts dealing with complex variables), it is found that

$$G = \frac{c}{R}\,\delta[R - c(t-\tau)], \tag{IX–25}$$

where $R^2 = (x_i - \xi_i)(x_i - \xi_i)$. Note that $G(x,t;\xi,\tau) = G(\xi,-\tau;x,-t)$. This Green's function for the wave equation represents an expanding concentrated wave. Where the domain has boundaries, it may be useful to adjoin to G as given by Eq. (IX–25) a "nonsingular" part; for example, the method of images could be used to construct the appropriate Green's function for a half-space.

Because of symmetry, G satisfies the equation

$$\nabla_0^2 G(x,t;\xi,\tau) - \frac{1}{c^2}\frac{\partial^2}{\partial\tau^2}G(x,t;\xi,\tau)$$
$$= -4\pi\,\delta(x_1-\xi_1)\,\delta(x_2-\xi_2)\,\delta(x_3-\xi_3)\,\delta(t-\tau), \tag{IX–26}$$

where $\nabla_0^2 = \partial^2/\partial\xi_i\partial\xi_i$. If now the function ϕ satisfies

$$\nabla_0^2\phi(\xi,\tau) - \frac{1}{c^2}\frac{\partial^2}{\partial\tau^2}\phi(\xi,\tau) = f(\xi,\tau), \tag{IX–27}$$

then multiplying Eq. (IX–26) by ϕ, (IX–27) by G, subtracting, integrating over whatever spatial volume is involved, using the divergence theorem, and integrating in time from $\tau = 0$ to $\tau = t+$ gives the formal solution of Eq. (IX–27) as

$$4\pi\phi(x,t) = \int_0^{t+} d\tau \left\{ -\int G(x,t;\xi,\tau)f(\xi,\tau)\,dV_0 \right.$$
$$\left. + \int \left[G(x,t;\xi,\tau)\frac{\partial\phi(\xi,\tau)}{\partial n_0} - \phi(\xi,\tau)\frac{\partial}{\partial n_0}G(x,t;\xi,\tau)\right] dS_0 \right\}$$
$$-\frac{1}{c^2}\int\left[\phi(\xi,\tau)\frac{\partial}{\partial\tau}G(x,t;\xi,\tau) - G(x,t;\xi,\tau)\frac{\partial}{\partial\tau}\phi(\xi,\tau)\right]_{\tau=0} dV_0. \tag{IX–28}$$

The subscript 0 refers to the ξ_i variables. As a check, the reader may use Eq. (IX–28) to rederive Eq. (IX–22).

Two special solutions are of some interest. For the case of an induced sinusoidal disturbance,

$$f = \delta(x_1)\,\delta(x_2)\,\delta(x_3)\cos\omega t,$$

Eq. (IX–28) gives

$$4\pi\phi = -\frac{C}{R}\cos\left[\omega(t - R/c)\right],$$

where R is the distance from the origin. (The easiest way to obtain this result is to let $\phi = \phi_t = 0$ at $t = 0$, and then take large t.) If, secondly, the disturbance moves at constant velocity, so that

$$f = \delta(x_1)\ \delta(x_2)\ \delta(x_3 - Vt),$$

then Eq. (IX–28) gives

$$4\pi\phi = -\frac{1}{[x_1^2 + x_2^2 + (x_3 - Vt')^2]^{\frac{1}{2}} - (V/c)(x_3 - Vt')},$$

where t' is the solution of

$$[x_1^2 + x_2^2 + (x_3 - Vt')^2]^{\frac{1}{2}} = c(t - t')$$

and thus has an obvious physical interpretation.

If it is known from physical considerations that the motion is steady and sinusoidal with fixed frequency ω, then the solution of Eq. (IX–7) may be written

$$\phi = \alpha \cos \omega t + \beta \sin \omega t,$$

where α and β are functions of the space coordinates alone; more compactly,

$$\phi = \mathrm{Re}(Ae^{i\omega t}),$$

where A is a complex spatial function. Equation (IX–7) then becomes

$$c^2 A_{,ii} + \omega^2 A = 0,$$

which is known as Helmholtz's equation. This equation must hold for both the real and imaginary parts of A. Green's-function methods may be used as before; since Helmholtz's equation is a special case of the wave equation, the corresponding Green's functions must of course be related.

IX–4. General Solutions of the Elastic-Wave Equation

The problem of writing a compact general solution for the elastic-wave equation corresponding to arbitrary initial and boundary conditions is tractable only for regions which are either infinite or semi-infinite; the basic reason for success in these cases is the availability of the powerful method of Fourier transforms. Transform methods have already been mentioned in Sec. IX–3; they could also have been employed in a very direct manner to obtain the concentrated-force solutions of Chapter V. For the elastic-wave equation, transform methods are indispensable; unfortunately, the details are rather technical, so that only the results will be quoted here. It may, however, be remarked in connection with these methods that in the case of an infinite body, subjected, for example, to an internal sinusoidally oscillating force, the motion is not that which would result from a limiting situation in which a finite body grows infinitely large. The reason is that, in any finite body, waves are reflected from the boundaries and so set up standing waves; in an infinite body this cannot occur and so outgoing waves will ensue. This "radiation condition" is useful in choosing between possible paths of contour integration, although such ambiguities may often be avoided by appropriate choice of initial conditions (for example, the use of Laplace rather than Fourier transforms in time).

Consider first the case of a disturbance originally confined to a region R of a large body (only if the solution is to hold for all time need the body be infinite). Let the value of u_i be desired at the point P at time t. Draw two spheres centered at P, S_1 of radius c_1t and S_2 of radius c_2t. Let V denote the portion of R contained between the two spheres. Let $u_i^{(0)}$ and $\dot{u}_i^{(0)}$ denote the displacement and velocity in R at time zero. Then [2]

$$\begin{aligned} u_i(P, t) = {} & \frac{1}{4\pi}\int (t\dot{u}_j^{(0)} + u_j^{(0)})\left(\frac{1}{r}\right)_{,ij} dV \\ & + \frac{1}{4\pi}\int \left\{ r\left[u_j^{(0)}\left(\frac{1}{r}\right)_{,ij}\right] + \left(\frac{1}{r}\right)_{,i}\left[t\dot{u}_j^{(0)} + u_j^{(0)} + r\frac{\partial u_j^{(0)}}{\partial r}\right] n_j \right\} dS_1 \\ & - \frac{1}{4\pi}\int \left\{ r\left[u_j^{(0)}\left(\frac{1}{r}\right)_{,ij}\right] + \left(\frac{1}{r}\right)_{,i}\left[t\dot{u}_j^{(0)} + u_j^{(0)} + r\frac{\partial u_j^{(0)}}{\partial r}\right] n_j \right. \\ & \left. - \frac{1}{r^2}\left[t\dot{u}_i^{(0)} + u_i^{(0)} + r\frac{\partial u_i^{(0)}}{\partial r}\right] \right\} dS_2, \end{aligned} \tag{IX–29}$$

where the outward normal from V is used in the surface integrals. Here r is the distance from P to the element of volume or area. It follows from Eq. (IX–29) that, if r_1 is the largest and r_2 the smallest distance from P to R, then the motion at P (if any) begins at time r_2/c_1 and ceases at time r_1/c_2.

Consider next the case of a concentrated body force in the x_3-direction of magnitude $\cos \omega t$ applied at the origin of an infinite body. After initial transients have died out, the displacement is

$$\begin{aligned} u_i = {} & \frac{\delta_{i3}}{4\pi G r}\cos\frac{\omega}{c_2}(c_2t - r) \\ & - \frac{c_2^2}{4\pi G\omega^2}\left\{\frac{1}{r}\left[\cos\frac{\omega}{c_1}(c_1t - r) - \cos\frac{\omega}{c_2}(c_2t - r)\right]\right\}_{,3i}, \end{aligned} \tag{IX–30}$$

[2] See A. E. H. Love, *London Math. Soc. Proc. 1* (1904), 291.

where G is the shear modulus, and r is the distance from the origin, and, since any forcing function may be regarded as a superposition of concentrated sinusoidal ones, this solution may be used to generate the displacement corresponding to any spatial and temporal force distribution. In particular, concentrated pressures and couples could be handled by differentiation, as in Sec. V–3. Equation (IX–30) is essentially due to Stokes.

If next a concentrated force of magnitude F directed along the positive x_3-axis moves with velocity $V < c_2$, then (disregarding initial transients) the displacement is:[3]

(*a*) for motion along the x_3-axis,

$$u_1 = \frac{x_1 F c_1^2}{4\pi G V^2 r^2}\left[\frac{\gamma_2}{(r^2+\gamma_2^2)^{\frac{1}{2}}} - \frac{\gamma_1}{(r^2+\gamma_1^2)^{\frac{1}{2}}}\right],$$

$$u_2 = \frac{x_2 F c_2^2}{4\pi G V^2 r^2}\left[\frac{\gamma_2}{(r^2+\gamma_2^2)^{\frac{1}{2}}} - \frac{\gamma_1}{(r^2+\gamma_1^2)^{\frac{1}{2}}}\right],$$

$$u_3 = -\frac{F c_2^2}{4\pi G V^2}\left[\frac{(1-V^2/c_2^2)^{\frac{1}{2}}}{(r^2+\gamma_2^2)^{\frac{1}{2}}} - \frac{(1-V^2/c_1^2)^{\frac{1}{2}}}{(r^2+\gamma_1^2)^{\frac{1}{2}}}\right],$$

where

$$\gamma_i = \frac{x_3 - Vt}{(1 - V^2/c_i^2)^{\frac{1}{2}}},$$

$$r^2 = x_1^2 + x_2^2;$$

(*b*) for motion along the x_1-axis,

$$u_1 = \frac{(x_1 - Vt)x_2 F c_2^2}{4\pi G V^2 r^2}\left(\frac{1}{R_2} - \frac{1}{R_1}\right),$$

$$u_2 = -\frac{x_2 x_3 F c_2^2}{4\pi G V^2 r^4}\left[R_2 - R_1 + (x_1 - Vt)^2\left(\frac{1}{R_2} - \frac{1}{R_1}\right)\right],$$

$$u_3 = \frac{F c_2^2}{4\pi G V^2}\left[\frac{V^2}{c_2^2 R_2} + \frac{x_2^2}{r^4}(R_2 - R_1) - \frac{(x_1 - Vt)^2 x_3^2}{r^4}\left(\frac{1}{R_2} - \frac{1}{R_1}\right)\right],$$

where

$$r^2 = x_2^2 + x_3^2,$$

$$R_i^2 = \left(1 - \frac{V^2}{c_i^2}\right)(x_2^2 + x_3^2) + (x_1 - Vt)^2.$$

[3] See G. Eason, J. Fulton, and I. Sneddon, *Phil. Trans. Roy. Soc.* (*London*) (*A*) *248* (1956), 575.

For a half-space, the evaluation of the resultant transform integrals is tedious, and reasonably simple results have been obtained for only a few cases. The most important of these is the effect of a concentrated pressure applied impulsively to the plane surface of the region $x_3 > 0$; Lamb [4] showed that on the surface the resulting dilatational and rotational disturbances are eventually dominated by slower and less rapidly attenuating surface waves of a kind previously studied by Rayleigh (see Sec. IX–5). Lamb [5] considers also the case of a moving pressure spot.

Equation (IX–30) may be used to generate the solution corresponding to a unit impulse; this solution may then be used as a Green's function exactly as in Eq. (IX–28). If boundaries are involved, then it may be necessary to adjoin to the Green's function a nonsingular part so chosen as to eliminate from the right-hand side of the equivalent of Eq. (IX–28) those terms involving unknown functions of displacement (just as in potential theory; see Sec. VI–1). However, for most of those cases where it is feasible to construct such a modified Green's function, the same method (for instance, transforms) that was used in its construction can be used directly to solve the problem. A possible exception occurs where the method of images can be used.

It was shown by Jacovache that the Galerkin solution (Sec. VI–4*f*) of the equations of static elasticity could be extended so as to generate solutions of the dynamic equations also. Sternberg and Eubanks subsequently proved that every displacement vector satisfying the dynamic equations could be expressed in the Jacovache form, and showed also that a simple transformation of the result yields an extension of the Papkovich-Neuber solution (Sec. VI–4*e*).

As in Sec. VI–4, assume that the effects of body forces have been removed by use of the previous special solutions, and write (see Sec. IX–1)

$$u_i = \phi_{,i} + e_{ijk}C_{k,j},$$

where

$$c_1^2\phi_{,jj} = \frac{\partial^2\phi}{\partial t^2}, \quad c_2^2C_{i,jj} = \frac{\partial^2 C_i}{\partial t^2}.$$

Define

$$P_i = \frac{1}{2(1-\sigma)}\left(\frac{c_1^2}{c_2^2}\,\phi_{,i}' + e_{ijk}C_{k,j}'\right), \tag{IX–31}$$

[4] H. Lamb, *Phil. Trans. Roy. Soc. (London) (A)* *203* (1904), 1.

[5] H. Lamb, *Phil. Mag.* (*6*) *13* (1916), 539.

where

$$\phi_{,jj}' - \frac{1}{c_2^2}\frac{\partial^2 \phi'}{\partial t^2} = \phi,$$

$$C_{i,jj}' - \frac{1}{c_1^2}\frac{\partial^2 C_i'}{\partial t} = C_i.$$

Incidentally, quantities defined in this manner may always be exhibited explicitly by use of Eqs. (IX–25) and (IX–28); for example,

$$\phi'(x, t) = -\frac{1}{4\pi}\int \frac{\phi(\xi, t - R/c_2)}{R}\, dV_0.$$

The vector P_i clearly satisfies

$$\left(\nabla^2 - \frac{1}{c_1^2}\frac{\partial^2}{\partial t^2}\right)\left(\nabla^2 - \frac{1}{c_2^2}\frac{\partial^2}{\partial t^2}\right) P_i = 0. \qquad \text{(IX–32)}$$

From Eq. (IX–31),

$$P_{i,i} = \frac{1}{2(1-\sigma)}\frac{c_1^2}{c_2^2}\phi_{,ii}' \qquad \text{(IX–33)}$$

and

$$\left(\nabla^2 - \frac{1}{c_1^2}\frac{\partial^2}{\partial t^2}\right) P_i = \frac{1}{2(1-\sigma)}\left[\frac{c_1^2}{c_2^2}\left(\nabla^2 - \frac{1}{c_1^2}\frac{\partial^2}{\partial t^2}\right)\phi_{,i}' + e_{ijk}C_{k,j}\right].$$

Using this last result to give an expression for $e_{ijk}C_{k,j}$,

$$u_i = \phi_{,i} + 2(1-\sigma)\left(\nabla^2 - \frac{1}{c_1^2}\frac{\partial^2}{\partial t^2}\right) P_i - \frac{c_1^2}{c_2^2}\left(\nabla^2 - \frac{1}{c_1^2}\frac{\partial^2}{\partial t^2}\right)\phi_{,i}',$$

and expressing ϕ in terms of ϕ' gives

$$u_i = 2(1-\sigma)\left(\nabla^2 - \frac{1}{c_1^2}\frac{\partial^2}{\partial t^2}\right) P_i + \left(1 - \frac{c_1^2}{c_2^2}\right)\phi_{,jji}'$$

$$= -P_{t,ti} + 2(1-\sigma)\left(\nabla^2 - \frac{1}{c_1^2}\frac{\partial^2}{\partial t^2}\right) P_i \qquad \text{(IX–34)}$$

by use of Eq. (IX–33). This may be compared with Eq. (VI–31). It is seen that the operator $\nabla^2 - (1/c_1^2)\partial^2/\partial t^2$ has replaced ∇^2; also, P_i satisfies Eq. (IX–32) rather than the condition of biharmonicity. The reader may prove the interesting fact that if $c_1 \neq c_2$ any quantity P_i satisfying

Eq. (IX–32) can always be written as the sum of P_i' and P_i'', where

$$\left(\nabla^2 - \frac{1}{c_1^2}\frac{\partial^2}{\partial t^2}\right) P_i' = 0,$$

$$\left(\nabla^2 - \frac{1}{c_2^2}\frac{\partial^2}{\partial t^2}\right) P_i'' = 0.$$

To extend now the Papkovich-Neuber representation, define

$$\eta_i = 2(1 - \sigma)\left(\nabla^2 - \frac{1}{c_1^2}\frac{\partial^2}{\partial t^2}\right) P_i,$$

$$\omega = -P_{j,j} + \frac{1}{4(1 - \sigma)} x_j\eta_j.$$

Then Eq. (IX–34) gives (see Eq. (VI–29))

$$u_i = -\frac{1}{4 - 4\sigma}(x_j\eta_j)_{,i} + \eta_i + \omega_{,i}, \qquad \text{(IX–35)}$$

where the functions η_i and ω, instead of being harmonic, satisfy

$$\left(\nabla^2 - \frac{1}{c_2^2}\frac{\partial^2}{\partial t^2}\right) \eta_i = 0,$$

$$\left(\nabla^2 - \frac{1}{c_1^2}\frac{\partial^2}{\partial t^2}\right) \omega = \frac{1}{4(1 - \sigma)} x_j\left(\nabla^2 - \frac{1}{c_1^2}\frac{\partial^2}{\partial t^2}\right) \eta_j.$$

IX–5. Surface Waves

It was remarked in Sec. IX–4 that disturbances at or near the surface of an elastic half-space are felt some distance away primarily in the form of special waves first discussed by Rayleigh in connection with earthquake phenomena. Consider a half-space in which the x_3-axis points into the material; the x_1- and x_2-axes lie in its surface. Let us try to construct a solution of the elastic-wave equations which satisfies the boundary conditions of zero τ_{13}, τ_{23}, τ_{33}, and which has the character of a plane wave of frequency ω proceeding with velocity c in the x_1-direction. Let $u_2 = 0$.

From Sec. IX–1, the displacement may be written in the form

$$u_i = \phi_{,i} + e_{ijk}C_{k,j},$$

where each of ϕ, C_i satisfies a simple wave equation. Since u_2 is to be assumed zero, ϕ depends only on x_1 and x_3; also, the second term, having only two nonvanishing components and zero divergence, may be written in terms of derivatives of a function $\psi(x_1, x_3)$. Thus

$$u_1 = \phi_{,1} + \psi_{,3}, \qquad u_3 = \phi_{,3} - \psi_{,1}, \tag{IX–36}$$

where

$$c_1^2 \nabla^2 \phi = \frac{\partial^2 \phi}{\partial t^2}. \tag{IX–37}$$

Use of the elastic-wave equation shows that ψ must satisfy

$$c_2^2 \nabla^2 \psi - \frac{\partial^2 \psi}{\partial t^2} = \text{some } f(t),$$

so that by adjoining to ψ a function of time whose second derivative cancels $f(t)$ (and doing this does not affect u_1 or u_3) we obtain a new ψ satisfying

$$c_2^2 \nabla^2 \psi - \frac{\partial^2 \psi}{\partial t^2} = 0. \tag{IX–38}$$

Consequently it is always possible to represent the wave (for which $u_2 = 0$) in the form of Eq. (IX–36), where Eqs. (IX–37) and (IX–38) are satisfied.

Assume now that ϕ has the form

$$\phi = A_1 \cos \omega \left(\frac{x_1}{c} - t \right) g_1(x_3).$$

Then Eq. (IX–37) requires that

$$g_1'' - p^2 g_1 = 0,$$

where

$$p^2 = \frac{\omega^2}{c^2} - \frac{\omega^2}{c_1^2}, \tag{IX–39}$$

which is tentatively assumed positive, in order to make possible a solution

$$g_1 = e^{-p x_3},$$

which diminishes rapidly for increasing x_3. Similarly, writing [6]

$$\psi = \left[B_1 \cos \omega \left(\frac{x_1}{c} - t \right) + B_2 \sin \omega \left(\frac{x_1}{c} - t \right) \right] g_2(x_3)$$

[6] It is necessary to allow for a possible phase difference between the ϕ and ψ waves.

and using Eq. (IX–38) shows that

$$g_2 = e^{-qx_3},$$

where

$$q^2 = \frac{\omega^2}{c^2} - \frac{\omega^2}{c_2^2}.$$

The boundary conditions at $x_3 = 0$,

$$\begin{aligned} u_{1,3} + u_{3,1} &= 0, \\ u_{2,3} + u_{3,2} &= 0, \\ (1-\sigma)u_{3,3} + \sigma u_{1,1} &= 0, \end{aligned}$$

then require $B_1 = 0$ and also

$$\left(2p\,\frac{\omega}{c}\right) A_1 + \left(q^2 + \frac{\omega^2}{c^2}\right) B_2 = 0,$$

$$\left[(1-\sigma)p^2 - \sigma\,\frac{\omega^2}{c^2}\right] A_1 + \left[(1-2\sigma)q\,\frac{\omega}{c}\right] B_2 = 0,$$

from which c may be found as the condition of vanishing of the coefficient determinant:

$$\left(1 - \frac{c^2}{2c_2^2}\right)^2 = \left(1 - \frac{c^2}{c_1^2}\right)^{\frac{1}{2}}\left(1 - \frac{c^2}{c_2^2}\right)^{\frac{1}{2}}. \qquad \text{(IX–40)}$$

Upon reduction, this is a cubic in c^2. For $\sigma = \frac{1}{4}$, the value of c is about 15 percent less than c_2. The ratio of A_1 to B_2 and the values of p and q are easily found once Eq. (IX–40) is solved. Note that the transverse and longitudinal displacements are exactly out of phase, that the velocity c is independent of ω, and that the depth of penetration does depend on ω. For $\sigma = \frac{1}{4}$, the transverse component will have decreased to $\frac{1}{3}$ of its surface value at a depth of about $2c_2/\omega$. Although only sinusoidal waves have been considered, the fact that the Fourier integral theorem allows us to write any wave shape as a superposition of sinusoidal ones means that there is no lack of generality. It is important to notice that when this Fourier superposition is carried out, a wave which is sharp on the surface is not sharp in the interior (because of the exponential ω-dependent multipliers of the sinusoidal terms). That it is not possible for a Rayleigh wave to vanish (for all x_3) outside a certain interval of $x_1 - ct$ follows also from the fact that, if it were possible, a disturbance would be propagating with a velocity different from c_1 or c_2, which is contrary to previous results.

Another general problem which arises at the surface of an elastic body is exemplified by the reflection of a plane transverse wave. Again, consider the half-space $x_3 > 0$, and let a wave incident on the surface from the interior be given by

$$u_1 = A \cos\alpha \cos\left[\omega\left(\frac{x_1 \sin\alpha - x_3\cos\alpha}{c_2} - t\right)\right],$$
$$u_2 = 0,$$
$$u_3 = A \sin\alpha \cos\left[\omega\left(\frac{x_1 \sin\alpha - x_3\cos\alpha}{c_2} - t\right)\right],$$

where α is the angle (less than 90°) between the normal to the wave front and the x_3-axis. This solenoidal wave is "polarized" in the (x_1, x_3) plane. It is not possible to superpose an optically reflected transverse wave

$$\begin{Bmatrix} u_1 \\ u_2 \\ u_3 \end{Bmatrix} = \begin{Bmatrix} \cos\alpha \\ 0 \\ -\sin\alpha \end{Bmatrix} \left\{ B_1 \cos\left[\omega\left(\frac{x_1\sin\alpha + x_3\cos\alpha}{c_2} - t\right)\right] + B_2 \sin\left[\omega\left(\frac{x_1\sin\alpha + x_3\cos\alpha}{c_2}\right) - t\right]\right\}$$

in such a way as to satisfy the boundary conditions of zero boundary stress. If, however, a "reflected" longitudinal wave is also allowed,

$$\begin{Bmatrix} u_1 \\ u_2 \\ u_3 \end{Bmatrix} = \begin{Bmatrix} \sin\beta \\ 0 \\ \cos\beta \end{Bmatrix} \left\{ C_1 \cos\left[\omega\left(\frac{x_1\sin\beta + x_3\cos\beta}{c_1} - t\right)\right] + C_2 \sin\left[\omega\left(\frac{x_1\sin\beta + x_3\cos\beta}{c_1} - t\right)\right]\right\},$$

where the angle of reflection β may differ from α, then choosing

$$\frac{\sin\beta}{c_1} = \frac{\sin\alpha}{c_2},$$
$$B_2 = C_2 = 0,$$
$$\frac{B_1}{A} = \frac{\cos^2 2\alpha - (c_2/c_1)^2 \sin 2\alpha \sin 2\beta}{\cos^2 2\alpha + (c_2/c_1)^2 \sin 2\alpha \sin 2\beta},$$
$$\frac{C_1}{A} = \frac{(c_2/c_1)\sin 4\alpha}{\cos^2 2\alpha + (c_2/c_1)^2 \sin 2\alpha \sin 2\beta}$$

allows satisfaction of the boundary conditions. Thus a vertically polarized incident transverse wave is reflected as a vertically polarized transverse wave plus a longitudinal wave; this latter leaves at an angle whose sine is related to the sine of the angle of incidence by a formula reminiscent of Snell's law in optics. Similarly, an incident longitudinal wave is reflected as the sum of a longitudinal wave and a vertically polarized transverse wave; however, an incident horizontally polarized wave is reflected without an accompanying longitudinal wave.

If two different elastic media are in contact, then both reflection and refraction may occur. The problem may be treated in the same general way; a thorough discussion will be found in reference 7.

IX–6. Vibrations

Most vibration problems in elasticity involve bodies for which various simplifications can be made; for example, in plates, line elements initially perpendicular to the neutral surface are assumed to remain so during the motion. Following such simplifications, the resultant differential equations may be handled by standard eigenvalue techniques. However, there are certain results concerning vibrations which may be discussed within the framework of general elastic bodies. For brevity of analysis, body forces will be omitted.

An elastic body is said to be vibrating in a *normal mode* if the motion is such that all particles are moving synchronously (and so passing through their "rest" positions simultaneously). Analytically,

$$u_i = A_i(x_1, x_2, x_3)f(t) \tag{IX–41}$$

At first sight, this type of motion seems artificial. However, it is easily shown [8] that for small motions of a body with a finite number of degrees of freedom it is possible to find a set of coordinates with the property that the differential equations of motion when expressed in these coordinates reduce to a set of separated equations each of which when integrated requires the coordinate in question to perform sinusoidal motion. The equations being separated, each such "normal mode" of oscillation is uncoupled from each other mode; further, since the set of "normal coordinates" is linearly related to the original coordinates (and vice versa),

[7] W. Ewing, W. Jardetzky, and F. Press, *Elastic waves in layered media* (McGraw-Hill, New York, 1957).

[8] R. Courant and D. Hilbert, *Methods of mathematical physics* (Interscience, New York, 1953), vol. 1, p. 281.

it follows that the most general motion of the body is a linear superposition of these normal modes, and that each mode if acting alone has the property of synchronous motion. Equation (IX–41) is a natural extrapolation of normal-mode motion to the case of a continuum. Substitution of Eq. (IX–41) into Eq. (IX–1) gives

$$G\left(A_{i,jj} + \frac{1}{1-2\sigma} A_{j,ji}\right) f = \rho A_i f''. \tag{IX–42}$$

Dividing by $A_i f$ and examining the resultant equation — one side of which is a function of (x_i) alone and the other of t alone — shows that f''/f can at most be a constant, say $-\omega^2$. Thus

$$f = \alpha \cos \omega t + \beta \sin \omega t,$$

and since the zero of time can be chosen arbitrarily we will drop the second term. Had not a real negative constant been chosen, the motion would have involved exponential growth or decay, and this is not the kind of motion in which we are presently interested. Then (absorbing α into A_i)

$$u_i = A_i \cos \omega t$$

and Eq. (IX–42) becomes

$$G\left(u_{i,jj} + \frac{1}{1-2\sigma} u_{j,ji}\right) = -\rho\omega^2 u_i. \tag{IX–43}$$

Certain boundary conditions are prescribed for the surface of the body, and it will in general turn out that these can be met only for certain values of ω, called characteristic values or *eigenvalues*. The corresponding u_i are called *eigenfunctions*. An analogous situation is that in which the equation

$$\frac{d^2y}{dx^2} + \lambda^2 y = 0$$

is required to have a nontrivial solution vanishing at $x = 0$ and $x = 1$; only for certain values of λ is this possible. For Eq. (IX–43), the boundary conditions met with in vibration problems are usually such that if v_i, w_i are any two displacement vectors satisfying the boundary conditions, then

$$\int T_i^{(v)} w_i \, dS = \int T_i^{(w)} v_i \, dS. \tag{IX–44}$$

For example, the surface may be unloaded everywhere except at certain regions of support where displacement is not allowed. For boundary conditions satisfying Eq. (IX–44), it is easy to prove that if $u_i^{(1)}$ is a solution corresponding to the eigenvalue ω_1, and $u_i^{(2)}$ is a solution corresponding to a different eigenvalue ω_2, then $u_i^{(1)}$ and $u_i^{(2)}$ are orthogonal (in a sense reminiscent of the function-space discussion of Sec. VI–6).

Multiplying both sides of Eq. (IX–43), as written for $u_i^{(1)}$, by $u_i^{(2)}$, and integrating over the volume of the body gives

$$G \int u_i^{(2)} \left(u_{i,jj}^{(1)} + \frac{1}{1 - 2\sigma} u_{j,ji}^{(1)} \right) dV = - \rho\omega_1^2 \int u_i^{(1)} u_i^{(2)} \, dV,$$

that is,

$$\int u_i^{(2)} \tau_{ij,j}^{(1)} \, dV = - \rho\omega_1^2 \int u_i^{(1)} u_i^{(2)} \, dV,$$

or, by the divergence theorem,

$$\int u_i^{(2)} T_i^{(1)} \, dS - \int u_{i,j}^{(2)} \tau_{ij}^{(1)} \, dV = - \rho\omega_1^2 \int u_i^{(1)} u_i^{(2)} \, dV. \quad \text{(IX–45)}$$

The second term on the left-hand side may be written

$$- \int e_{ij}^{(2)} \tau_{ij}^{(1)} \, dV = - \frac{1}{E} \int [(1 + \sigma) \tau_{ij}^{(2)} \tau_{ij}^{(1)} - \sigma \tau_{kk}^{(1)} \tau_{jj}^{(2)}] \, dV,$$

which is symmetric in the superscripts (1) and (2). Repeating the calculation of Eq. (IX–45) with the roles of the superscripts reversed, and using Eq. (IX–44), it follows that

$$(\omega_1^2 - \omega_2^2) \int u_i^{(1)} u_i^{(2)} \, dV = 0,$$

so that, because of the fact that ω_1^2 differs from ω_2^2,

$$\int u_i^{(1)} u_i^{(2)} \, dV = 0, \quad \text{(IX–46)}$$

which is the required orthogonality condition. Incidentally, this kind of proof can easily be used to show also that the ω_i are real. It is merely necessary to note that, if ω_1, $u_i^{(1)}$ are complex, then their conjugates must represent different solutions which could be denoted by ω_2, $u_i^{(2)}$; the result (IX–46) would then be replaced by

$$\int u_i^{(1)} (u_i^{(1)})^* \, dV = 0$$

(where the star represents complex conjugate) — an impossible result since the integrand is nonnegative.

By use of the Green's-function methods of Chapter VI, Eq. (IX–43) (with appropriate boundary conditions) can be put in the form of a homogeneous integral equation with symmetric kernel; for such equations, it is known that there is at least one eigenvalue, that the eigenvalues are all denumerable (that is, they do not cluster), and that a repeated eigenvalue of multiplicity n has n linearly independent eigenfunctions u_i associated with it. Beyond providing these general results, the integral-equation formulation may allow convenient numerical methods to be used for finding eigenvalues (or bounds on them) and eigenfunctions; however, these topics are adequately treated elsewhere.

Returning to Eq. (IX–43) and writing it in the form

$$L_i(u_1, u_2, u_3) = -\rho\omega^2 u_i,$$

we obtain upon multiplication by u_i and integration,

$$\omega^2 = -\frac{\int u_i L_i \, dV}{\rho \int u_i u_i \, dV}. \tag{IX–47}$$

This type of right-hand-side expression is referred to as "Rayleigh's Quotient." It is easily shown by variational methods (see Chapter VII) that the stationary values of the quotient correspond to eigenfunctions. Consequently, one way of determining approximately the eigen quantities is to guess mode shapes in forms involving a number of arbitrary constants which are then chosen by the Rayleigh-Ritz method (Chapter VII); it may be convenient to insist that each mode be orthogonal to any lower ones previously found. (Since ω^2 must be positive, the sign of the numerator of Eq. (IX–47) must be negative, a fact which follows also from the positive nature of the strain energy.) There are many ways of approximating or bounding eigenvalues,[9] but these need not be discussed here.

In general, there are two reasons for one's interest in eigenfrequencies. First, considering a continuum as the limiting case of a system with a finite number of degrees of freedom as discussed earlier in this section, it is clear that any vibratory motion of the body must be a superposition of natural modes. Second, these modes involve the natural frequencies at which external forces could act so as to cause vibratory destruction.

[9] L. Collatz, *Eigenwertprobleme* (Chelsea, New York, 1948).

X

Nonlinear Elasticity

X–1. Stress-Strain Law

If a body is moving under the action of body and surface forces, the rate at which these forces do work on the body is

$$\dot{W} = \int \rho F_i \dot{u}_i \, dV + \int T_i \dot{u}_i \, dS,$$

which easily reduces (see Sec. V–7) to

$$\dot{W} = \dot{K} + \int \tau_{ij} \tfrac{1}{2}(\dot{u}_{i,j} + \dot{u}_{j,i}) \, dV$$

where $\dot{K}$ is the rate of increase of kinetic energy. By use of Eq. (IV–20), this result becomes

$$\dot{W} = \dot{K} + \int \tau_{ij} \dot{\eta}_{pq} \frac{\partial a_p}{\partial x_i} \frac{\partial a_q}{\partial x_j} \, dV. \qquad \text{(X–1)}$$

This result is exact. Define now the *elastic range* as that range of states which can be reached reversibly (in the thermodynamic sense of Chapter VIII) from the unloaded constant-temperature state. Each such state has associated with it a certain value of entropy S, internal energy U, and free energy $F = U - TS$.

Equation (X–1) is an identity between mechanical quantities. In order to make use of thermodynamics, it is necessary to specify whether or not thermal exchanges are taking place between the body and its surroundings. Let us insist first of all that the motion described by Eq. (X–1) is to take place sufficiently slowly that $\dot{K}$ (which depends quadratically on $\dot{u}_i$) can be neglected, and also that the body is to be maintained at constant temperature T. Then the rate at which heat is being added to the body, for such a reversible process, is

$$T\dot{S} = \dot{U} - \dot{W},$$

that is, since $\dot{T} = 0$,

$$\dot{F} = \frac{d}{dt}(U - TS) = \dot{W}.$$

If ϕ is the free energy per unit mass — a function of T and η_{ij}, assumed written symmetrically in the latter — then this equation becomes

$$\int \rho\dot{\phi}\, dV = \int \tau_{ij}\dot{\eta}_{pq} \frac{\partial a_p}{\partial x_i}\frac{\partial a_q}{\partial x_j}\, dV.$$

Since this must hold for each portion of the body,

$$\rho\dot{\phi} = \tau_{ij}\dot{\eta}_{pq} \frac{\partial a_p}{\partial x_i}\frac{\partial a_q}{\partial x_j}. \tag{X–2}$$

Writing

$$\dot{\phi} = \frac{\partial \phi}{\partial \eta_{pq}}\dot{\eta}_{pq} + \frac{\partial \phi}{\partial T}(0),$$

and using the symmetry of ϕ, the respective coefficients of $\dot{\eta}_{pq}$ must be equal:

$$\tau_{ij} \frac{\partial a_p}{\partial x_i}\frac{\partial a_q}{\partial x_j} = \rho \frac{\partial \phi}{\partial \eta_{pq}}$$

or

$$\tau_{ij} = \rho \frac{\partial \phi}{\partial \eta_{pq}} \frac{\partial x_i}{\partial a_p}\frac{\partial x_j}{\partial a_q}, \tag{X–3}$$

a result first obtained by Murnaghan. Equation (X–3) gives the τ_{ij} in terms of η_{ij} and T; since the stresses must depend only on the state of the body and not on how it got there, Eq. (X–3) is universally valid irrespective of the fact that it was derived by considering the behavior of the body during an isothermal deformation.

Equivalently, a reversible adiabatic deformation could have been used. If so, $\dot{W} = \dot{U}$, and Eq. (X–3) would be replaced by

$$\tau_{ij} = \rho \frac{\partial u}{\partial \eta_{pq}} \frac{\partial x_i}{\partial a_p}\frac{\partial x_j}{\partial a_q}, \tag{X–4}$$

where the internal energy per unit mass, u, is considered written symmetrically in terms of the η_{ij}. There is however an important difference between the partial derivatives $\partial\phi/\partial\eta_{pq}$ and $\partial u/\partial\eta_{pq}$ in that T is constant in the former and s, the entropy per unit mass, is constant in the latter (and thus u is considered to be a function of η_{ij} and s in Eq. (X–4)). The fact that the results must be identical means that, for the same values of whatever variables are used to specify the state,

$$\left(\frac{\partial \phi}{\partial \eta_{pq}}\right)_T = \left(\frac{\partial u}{\partial \eta_{pq}}\right)_s. \quad \text{(X–5)}$$

Equation (X–5) is analogous to the result

$$\left(\frac{\partial F}{\partial V}\right)_T = \left(\frac{\partial U}{\partial V}\right)_S,$$

which holds for the case of a simple substance (Sec. VIII–2), and could in fact be derived in the same way.

The strain has been expressed in terms of Lagrangian coordinates in Eq. (X–3) for the reason that ϕ is not in general a function of (e_{ij}, T). If, however, the body is isotropic, so that orientation is unimportant, then ϕ may alternatively be written as $\phi(e_{ij}, T)$ and an equation similar to Eq. (X–3) can be derived (Sec. X–2). For the present, we merely note the interesting relation derivable from Eqs. (X–3) and (IV–17):

$$e_{ij}\tau_{ij} = \rho \frac{\partial \phi}{\partial \eta_{pq}} \eta_{pq}. \quad \text{(X–6)}$$

Most applications of Eqs. (X–3) or (X–4) involve the expansion of ϕ or u in power series in the η_{ij}. For example, if

$$\phi = \phi_0 + A_{ij}\eta_{ij} + B_{ijkl}\eta_{ij}\eta_{kl} + \cdots,$$

where $B_{ijkl} = B_{klij}$, then

$$\tau_{ij} = \rho \frac{\partial x_i}{\partial a_p}\frac{\partial x_j}{\partial a_q}(A_{pq} + 2B_{pqkl}\eta_{kl} + \cdots), \quad \text{(X–7)}$$

and, if τ_{ij} is to vanish for zero deformation, $A_{pq} = 0$. If desired, we could write

$$\frac{\partial x_i}{\partial a_p} = \delta_{ip} + \frac{\partial u_i}{\partial a_p}$$
$$= \delta_{ip} + \eta_{ip} + \gamma_{ip},$$

where γ_{ip} is the rotation tensor of Chapter IV, and use Eq. (IV–28), so as to obtain Eq. (X–7) in terms of η_{ij} and γ_{ij} alone.

X–2. Eulerian Forms

Even though ϕ cannot (for nonisotropic bodies) be expressed in terms of e_{ij} rather than η_{ij}, an Eulerian type of expression of Eq. (X–3) is still possible. Define (cf. Eqs. (IV–14) and (IV–15))

$$j_{ij} = \frac{\partial a_i}{\partial x_s}\frac{\partial a_j}{\partial x_s},$$

so that

$$j_{ij}(2\eta_{sj} + \delta_{sj}) = \delta_{is}. \quad \text{(X–8)}$$

For any choice of s, this is a set of three linear equations which may be solved for the η_{ij} in terms of the j_{ij}. (In fact,

$$\eta_{ij} = \frac{1}{4D} e_{ikp} e_{jst} j_{ks} j_{pt} - \frac{1}{2} \delta_{ij},$$

where D is the determinant $|j_{ij}|$.) However, all we need is that each η_{ij} is some function of the j_{ij}; it then follows that ϕ may be expressed as $\phi(j_{ij}, T)$.

Next, differentiate Eq. (X–8) with respect to η_{pq} to give

$$\frac{\partial j_{ij}}{\partial \eta_{pq}} (2\eta_{sj} + \delta_{sj}) + 2 j_{iq} \delta_{sp} = 0$$

and multiply by j_{sr} to give

$$\frac{\partial j_{ir}}{\partial \eta_{pq}} = -\, 2 j_{iq} j_{rp}. \tag{X–9}$$

Consequently, if ϕ is written symmetrically in the j_{ij}, Eq. (X–3) becomes

$$\begin{aligned} \tau_{ij} &= \rho \frac{\partial \phi}{\partial j_{sr}} (-\, 2 j_{sq} j_{rp}) \frac{\partial x_i}{\partial a_q} \frac{\partial x_j}{\partial a_p} \\ &= -\, 2\rho \frac{\partial \phi}{\partial j_{sr}} \frac{\partial a_s}{\partial x_i} \frac{\partial a_r}{\partial x_j}, \end{aligned} \tag{X–10}$$

which is the required result.

If the medium is isotropic, Murnaghan's stress-strain law (X–3) attains a particularly simple form. Write ϕ symmetrically in terms of (e_{ij}, T) — which is now possible, since a statement of the values of the e_{ij} alone is enough to specify completely the state of strain insofar as the energy per unit mass of an isotropic body is concerned. Returning to Eq. (X–2), $\dot{\phi}$ is now written

$$\begin{aligned} \dot{\phi} &= \frac{\partial \phi}{\partial e_{ij}} \dot{e}_{ij} \\ &= \frac{\partial \phi}{\partial e_{ij}} \frac{d}{dt} \left(-\frac{1}{2} \frac{\partial a_s}{\partial x_i} \frac{\partial a_s}{\partial x_j} \right) \end{aligned}$$

by Eq. (IV–15); using Eq. (IV–19),

$$\begin{aligned} \dot{\phi} &= \frac{1}{2} \frac{\partial \phi}{\partial e_{ij}} \left(\frac{\partial a_s}{\partial x_i} \frac{\partial a_s}{\partial x_t} \dot{u}_{t,j} + \frac{\partial a_s}{\partial x_j} \frac{\partial a_s}{\partial x_t} \dot{u}_{t,i} \right) \\ &= \frac{\partial \phi}{\partial e_{ij}} \frac{\partial a_s}{\partial x_i} \frac{\partial a_s}{\partial x_t} \dot{u}_{t,j} \end{aligned}$$

because of the symmetry of ϕ. Consequently Eq. (X–2) becomes

$$\tau_{tj}\dot{u}_{t,j} = \rho \frac{\partial \phi}{\partial e_{ij}} \frac{\partial a_s}{\partial x_i} \frac{\partial a_s}{\partial x_t} \dot{u}_{t,j}$$

$$= \rho \frac{\partial \phi}{\partial e_{ij}} (\delta_{it} - 2e_{it})\dot{u}_{t,j}.$$

This equation can hold for all velocity functions $\dot{u}_i$ only if the two coefficients of $\dot{u}_{t,j}$ are equal:

$$\tau_{ij} = \rho \frac{\partial \phi}{\partial e_{ij}} - 2\rho e_{is} \frac{\partial \phi}{\partial e_{sj}}. \qquad \text{(X–11)}$$

An interesting consequence of this equation is that the quantity

$$e_{is} \frac{\partial \phi}{\partial e_{sj}}$$

must be symmetric, since all other terms are. Thus

$$e_{is} \frac{\partial \phi}{\partial e_{sj}} = e_{js} \frac{\partial \phi}{\partial e_{si}}. \qquad \text{(X–12)}$$

Let us now prove that

$$\tau_{ij}e_{ip} = \tau_{ip}e_{ij}, \qquad \text{(X–13)}$$

so that by Sec. III–4 (6) it will follow that the stress and strain tensors have the same principal axes for even a nonlinear isotropic elastic body. The left-hand side is, by Eqs. (X–11) and (X–12),

$$\tau_{ij}e_{ip} = \rho \left(\frac{\partial \phi}{\partial e_{ij}} e_{ip} - 2e_{is} \frac{\partial \phi}{\partial e_{sj}} e_{ip} \right)$$

$$= \rho \left(\frac{\partial \phi}{\partial e_{ip}} e_{ij} - 2e_{js} \frac{\partial \phi}{\partial e_{si}} e_{ip} \right)$$

$$= \rho \left(\frac{\partial \phi}{\partial e_{ip}} e_{ij} - 2e_{ji} \frac{\partial \phi}{\partial e_{is}} e_{sp} \right)$$

(merely altering dummy indices)

$$= \rho \left(\frac{\partial \phi}{\partial e_{ip}} e_{ij} - 2e_{ij} \frac{\partial \phi}{\partial e_{sp}} e_{si} \right)$$

$$= \tau_{ip}e_{ij},$$

which is the required result.

The coincidence of principal axes has some interesting consequences. If a principal-axis system is chosen, Eq. (X–11) gives, for example,

$$\tau_{11} = \rho(1 - 2e_{11}) \frac{\partial \phi}{\partial e_{11}},$$

$$\tau_{12} = \rho(1 - 2e_{11}) \frac{\partial \phi}{\partial e_{12}}.$$

Since $\tau_{12} = 0$, the second equation requires that $\partial\phi/\partial e_{12}$ vanish when $e_{12} = e_{13} = e_{23} = 0$; this fact allows one to simplify the general functional representation of ϕ in, say, a power series. The first equation shows that, since all off-diagonal e_{ij} are to be set equal to zero after the differentiation anyway, we might as well do this initially and consider ϕ to be a function of e_{11}, e_{22}, e_{33} alone. Thus, as long as principal axes are to be used,

$$\phi = \phi(e_1, e_2, e_3),$$

where $e_1 = e_{11}$, and so forth, and

$$\tau_1 = \rho(1 - 2e_1) \frac{\partial \phi}{\partial e_1},$$

and so forth. A power-series expansion for ϕ now shows (using isotropy) that, if second-order terms are to be included, then three new elastic constants are required in addition to the two old ones. Geometrically, the last equation says that the quantities $\tau_i/\rho(1 - 2e_i)$ are the components of the gradient of ϕ in (e_i) space.

If it is not desired to use principal axes, then it may be noted that, since the state of strain is completely known (insofar as energy is concerned) by a specification of the three strain invariants E_i of Chapter IV, ϕ may be written as $\phi(E_1, E_2, E_3)$.

X–3. Partial Nonlinearity

In some nonlinear elasticity problems, the nonlinearity in one of (*a*) the relation between e_{ij} and u_i, (*b*) the stress-strain relation, may be assumed unimportant, so that the problem becomes one of only partial nonlinearity. One such situation arises when the displacements are large, but the deformations are small; then the complete nonlinear formula linking u_i and e_{ij} must be used (or at least an approximation to it, as in Sec. IV–3), but it may nevertheless be permissible to require a linear relation between τ_{ij} and e_{ij}. Another case is that in which the displacements are small, so that the usual linear approximation may be made in the expression for e_{ij} in terms of u_i, but where the material is elastically nonlinear as reflected in the stress-strain law.

If one of these conditions is met, then it may be useful to have variational principles available for use in constructing approximate solutions. The principle of virtual work, being a fundamental theorem of mechanics, must always hold; in general, however, there is no analogue in nonlinear elasticity to the principle of virtual stress. An exception occurs in the second case above, where such an analogue has been given by Greenberg.

Following Greenberg, let us consider the problem

$$\begin{aligned} \tau_{ij,j} &= 0 \text{ in } V, \\ e_{ij} &= \tfrac{1}{2}(u_{i,j} + u_{j,i}) \text{ in } V, \\ u_i &= u_i^0 \text{ on } S_A, \\ \tau_{ij}n_j &= T_i^0 \text{ on } S_B, \\ e_{ij} &= e_{ij}(\tau_{11}, \tau_{12}, \cdots), \\ \tau_{ij} &= \tau_{ij}(e_{11}, e_{12}, \cdots), \end{aligned}$$

where the stress-strain relation is possibly nonlinear. An internal elastic energy function ξ per unit volume is assumed to exist, so that

$$d\xi = \tau_{ij}\, de_{ij},$$

$$\xi = \int_0^{e_{ij}} \tau_{ij}(e_{11}, e_{12}, \cdots)\, de_{ij},$$

where the integral is independent of path of integration. Then first, the principle of virtual work — here simply an identity — says that if u_i is the solution, then

$$\delta\left(\int \xi\, dV - \int T_i^0 u_i\, dS_B\right) = 0$$

for all δu_i satisfying $\delta u_i = 0$ on S_A. Second, the following identity is easily verified; it corresponds to the previous principle of virtual stress. If τ_{ij} is a solution, then

$$\delta\left[\int (e_{ij}\tau_{ij} - \xi)\, dV - \int u_i^0 \tau_{ij} n_j\, dS_A\right] = 0$$

for all $\delta\tau_{ij}$ satisfying $\delta\tau_{ij,j} = 0$ in V and $\delta\tau_{ij}n_j = 0$ on S_B. Here e_{ij} and ξ are considered to be expressed in terms of τ_{ij}.

The converses of these theorems are also true — that is, if such u_i or τ_{ij}

are found, then they satisfy the elasticity differential equations — as is easily seen by methods similar to those used in proving the converses for the corresponding linear situations in Chapter VII.

X–4. Elastic Stability

The basic problem of elastic stability is to determine whether or not a particular equilibrium state of an elastic body is stable. The test for stability consists in considering the effect of small disturbances applied to whatever idealized mathematical model is representing the actual structure; if the disturbances tend to grow without limit, then it is reasonable to assume that the actual structure will be unsafe. Rather than consider the solutions of the differential equations governing such disturbances, we will use an energy criterion, which states that an elastic system is unstable if there exists a virtual displacement having the property that the corresponding work done by the loading exceeds the increment of strain energy. Had we instead considered an elastic system originally in stable equilibrium under certain loading, and asked for that value of the loading at which a geometrically adjacent equilibrium position first becomes possible, the same results would have been obtained. Again, the differential-equation method would also have agreed. These statements are, however, no longer true if the system is not conservative.[1]

In applying the present or any other criterion, the immediate problem is to choose the appropriate stress-strain and strain-displacement relations. That some sort of nonlinearity is required follows from the fact that motion away from an equilibrium state must involve second-order terms; also, it is desired to avoid the uniqueness[2] theorem of linear elasticity (although strictly speaking the proof of this theorem contemplates the same body configuration for the two supposedly different states to be proved identical, and so the theorem is not applicable to stability problems in any event). Included in these considerations must be the behavior of the loading itself during a small displacement; it will turn out, for example, that "pressure-type" loads (force always proportional to and perpendicular to area element) and "dead-type" loads (force always constant in magnitude and fixed in direction) lead to quite different results. A number of writers (Bryan, Southwell, Biezeno, Hencky, Trefftz, Biot, Neuber, Prager, Goodier, Plass, Novozhilov, and

[1] See H. Ziegler, *Z. angew. Math. u. Phys. 4* (1953), 89, 167.

[2] The relations between stability and uniqueness in the nonlinear case have been discussed by R. Hill, *J. Mech. Phys. Solids 5* (1957), 229.

others) have considered the general problem of elastic stability; because of the different nonlinearities assumed for the stress-strain law, the results of different authors do not in general coincide. We will here make use of the exact nonlinear stress-strain law (X–3).

Consider an arbitrary elastic body initially free from stress (state I). On the application of load or of heat, the body alters position and shape (state II). The material particle initially at (a_i) has moved to (x_i), in terms of a fixed Cartesian coordinate system. Define the displacement vector by $v_i = x_i - a_i$ (we reserve the notation u_i for additional displacements from state II). Then

$$\eta_{ij} = \frac{1}{2}\left[\frac{\partial v_i}{\partial a_j} + \frac{\partial v_j}{\partial a_i} + \left(\frac{\partial v_s}{\partial a_i}\right)\left(\frac{\partial v_s}{\partial a_j}\right)\right],$$

and

$$\tau_{ij} = \rho\left(\frac{\partial U}{\partial \eta_{pq}}\right)_S \left(\frac{\partial x_i}{\partial a_p}\right)\left(\frac{\partial x_j}{\partial a_q}\right),$$

where U is the internal energy per unit mass. It is now required to analyze the stability of the body in its deformed state II. The body force per unit mass, F_i, is assumed constant. The surface loading T_i in state II is assumed to be produced by loads which vary neither in direction nor magnitude during the trial displacement; thus, under such "dead loading" the material particles constituting a portion of the surface in state II will always experience the same total vector surface force, irrespective of their subsequent orientation or total area. The work done by T_i and F_i in a trial displacement u_i is (exactly)

$$W = \int \rho F_i u_i \, dV + \int T_i u_i \, dS,$$

where the volume and surface integrals are calculated for state II. Altering to volume integrals and using the equations of equilibrium gives

$$\begin{aligned} W &= \int \tau_{ij}\left(\frac{\partial u_i}{\partial x_j}\right) dV \\ &= \int \frac{\partial U}{\partial \eta_{pq}} \frac{\partial x_i}{\partial a_p} \frac{\partial u_i}{\partial a_q} \rho \, dV. \end{aligned}$$

The exact increase in internal energy is

$$\int (U' - U)\rho \, dV,$$

where U' denotes the internal energy per unit mass following the displacement u_i, and depends on temperature as well as on u_i. Note that it is allowable to continue to calculate this volume integral for state II because of the invariance of the element of mass $\rho\, dV$. The condition for stability is that, for each allowable u_i,

$$\int \left[\frac{\partial U}{\partial \eta_{pq}} \frac{\partial x_i}{\partial a_p} \frac{\partial u_i}{\partial a_q} - (U' - U)\right] \rho\, dV \leq 0. \qquad \text{(X–14)}$$

It is next necessary to calculate U'. Because buckling is rapid, it is reasonable to require the u_i motion to be adiabatic. (Had we here insisted on an isothermal motion, the final results would have involved the isothermal rather than the adiabatic elastic constants, and the differences are small. In practice, the motion is probably somewhere between these two extremes.) Then a power-series expansion gives

$$U' - U = \frac{\partial U}{\partial \eta_{pq}}\, \delta\eta_{pq} + \frac{1}{2} \frac{\partial^2 U}{\partial \eta_{pq} \partial \eta_{ij}}\, \delta\eta_{ij}\, \delta\eta_{pq},$$

where all partial derivatives are to be evaluated for constant entropy and for state II. Using

$$\delta\eta_{ij} = \frac{1}{2}\left(\frac{\partial x_r}{\partial a_i}\frac{\partial u_r}{\partial a_j} + \frac{\partial x_r}{\partial a_j}\frac{\partial u_r}{\partial a_i} + \frac{\partial u_r}{\partial a_i}\frac{\partial u_r}{\partial a_j}\right)$$

alters the stability condition (X–14) to read

$$\int \rho\, dV \left(\frac{\partial U}{\partial \eta_{ij}} \frac{\partial u_r}{\partial a_i}\frac{\partial u_r}{\partial a_j} + \frac{\partial^2 U}{\partial \eta_{ij} \partial \eta_{pq}}\, \delta\eta_{ij} \delta\eta_{pq} + \cdots\right) > 0$$

for all nonzero permissible u_i. Using the stress-strain law, and dropping terms of third and higher order, which presumably are of interest only in pathological cases, the stability criterion becomes

$$\int dV \left(\tau_{pq} \frac{\partial u_r}{\partial x_p}\frac{\partial u_r}{\partial x_q} + \rho \frac{\partial^2 U}{\partial \eta_{ij} \partial \eta_{pq}} \frac{\partial x_r}{\partial a_i}\frac{\partial x_s}{\partial a_p}\frac{\partial u_r}{\partial a_j}\frac{\partial u_s}{\partial a_q}\right) > 0. \qquad \text{(X–15)}$$

All quantities are calculated for state II. This criterion is exact, and must be used whenever nonlinearity of the stress-strain law is essential.

In usual engineering applications, it is reasonable to approximate the coefficient of the second term by its value in state I, that is,

$$\left(\rho \frac{\partial^2 U}{\partial \eta_{ij} \partial \eta_{pq}}\right)_{II} \cong \rho_0 \left(\frac{\partial^2 U}{\partial \eta_{ij} \partial \eta_{pq}}\right)_I,$$

because of the fact that the additional terms for the power-series expansion of the left-hand side (representing nonlinear effects) are usually smaller than the uncertainty in the first term. But this first term is the usual adiabatic elastic coefficient, and if the deformation (although not necessarily the displacement) of state II is also assumed small then the partial derivative of (x_i) with respect to (a_i) must represent primarily rotation which affects the elastic coefficients according to their tensor character. Finally, assuming an isotropic medium, Eq. (X–13) becomes in engineering approximation

$$\int dV \left[\tau_{pq} \frac{\partial u_r}{\partial x_p} \frac{\partial u_r}{\partial x_q} + 2G \left(e_{sm} e_{sm} + \frac{\sigma}{1 - 2\sigma} e_{tt} e_{mm} \right) \right] > 0, \qquad \text{(X–16)}$$

where

$$e_{ij} = \tfrac{1}{2}(u_{i,j} + u_{j,i}).$$

If the curvilinear coordinates represented by

$$ds^2 = h_1^2 \, dy_1^2 + h_2^2 \, dy_2^2 + h_3^2 \, dy_3^2$$

are to be used, then representing by τ_{ij} the curvilinear stress components and by u_i the curvilinear displacement component (that is, in the parametric direction of y_i), the first term of Eq. (X–16) becomes (temporarily abandoning the usual summation convention)

$$\sum_{p,q,r,m} \frac{\tau_{pq}}{h_p h_q} \left[u_{r,p} u_{r,q} + \frac{u_p u_q}{h_r^2} h_{p,r} h_{q,r} + \left(\frac{u_r u_m}{h_r h_m} \delta_{pq} h_{p,m} h_{p,r} + \frac{2u_m}{h_m} h_{q,m} u_{q,p} - \frac{2u_q}{h_r} u_{r,p} h_{q,r} - \frac{2u_m u_p}{h_m h_q} h_{p,q} h_{q,m} \right) \right],$$

where a comma indicates differentiation with respect to the y-coordinates. The form of the second term is unaltered, but e_{ij} must now be interpreted as a curvilinear strain component:

$$e_{ij} = \frac{1}{2} \sum_s \left[\frac{h_j}{h_i} \frac{\partial}{\partial y_i} \left(\frac{u_j}{h_j} \right) + \frac{h_i}{h_j} \frac{\partial}{\partial y_j} \left(\frac{u_i}{h_i} \right) + 2\delta_{ij} \frac{u_s}{h_s h_j} \frac{\partial h_i}{\partial y_s} \right].$$

Finally, it may be noted that forces exerted by fixed constraints are included in the foregoing theory, for such forces do not work in an allowable trial displacement. If, however, a force is of "pressure type," the work done by a pressure P on a surface portion S_p through u_i can be calculated (note that the system is still assumed to be conservative) by allowing the intermediate displacement to grow at a constant rate —

that is, if t is time, let the displacement at time t be $(u_i t)$ and calculate the work done from $t = 0$ to $t = 1$. This work, W_1, is

$$W_1 = \int_0^1 dt \int (dS_p)_t \left[\frac{d}{dt} (u_i t)(-P)(n_i)_t \right],$$

where the subscript t denotes evaluation at time t. But

$$(n_i)_t (dS_p)_t = \frac{1}{2} e_{ijk} e_{rpq} \frac{\partial(x_j + u_j t)}{\partial x_p} \frac{\partial(x_k + u_k t)}{\partial x_q} n_r \, dS_p,$$

so that calculation gives

$$W_1 = -P \int dS_p [n_i u_i + \tfrac{1}{2}(n_i u_{k,k} u_i - n_k u_{k,i} u_i) + \tfrac{1}{6} e_{ijk} e_{rpq} u_{j,p} u_{k,q} u_i].$$

The first term of this expression would already have been included if P were treated as a dead load; consequently, the additional work done is that due to the remaining terms. Again omitting third-order terms in u_i, and remembering that the factor -2 was incorporated into Eq. (X–13), the term that must be adjoined to Eq. (X–15) for pressure loading is

$$\int dS_p \, . \, P(n_i u_{k,k} u_i - n_k u_{k,i} u_i).$$

Finally, let us consider a simplification of Eq. (X–16) that is usually permissible for metals, where the stresses are small compared to the elastic moduli. Writing

$$\begin{aligned} \frac{\partial u_r}{\partial x_p} &= \frac{1}{2}\left(\frac{\partial u_r}{\partial x_p} + \frac{\partial u_p}{\partial x_r}\right) + \frac{1}{2}\left(\frac{\partial u_r}{\partial x_p} - \frac{\partial u_p}{\partial x_r}\right) \\ &= e_{rp} + \omega_{rp} \end{aligned}$$

makes the first term of Eq. (X–16) read

$$\tau_{pq}(e_{rp} e_{rq} + 2 e_{rq} \omega_{rp} + \omega_{rp} \omega_{rq}). \qquad \text{(X–17)}$$

Comparing this expression with the second term of Eq. (X–16) shows that, because of the disparity in size between τ_{pq} and G, the only important contributions of expression (X–17) must be those involving ω_{ij} — and of those, the last term in (X–17) dominates. Consequently, it is usually permissible to write Eq. (X–16) as

$$\int dV \left[\tau_{pq} \omega_{rp} \omega_{rq} + 2G \left(e_{sm} e_{sm} + \frac{\sigma}{1 - 2\sigma} e_{tt} e_{mm} \right) \right] > 0. \qquad \text{(X–18)}$$

X-5. Nominal Stress

Some writers on finite strain refer all quantities to the original unstressed position of the body. Thus the nominal stress tensor τ_{ij}^0 is defined so that $\tau_{ij}^0\, dS_0$ is the jth component of the force presently acting on an area element originally perpendicular to the x_i-axis and of original area dS_0. For an area element dS_0 originally perpendicular to n_i^0, the real stress vector has components $T_i^0\, dS_0$, where

$$T_i^0 = \tau_{ji}^0 n_j^0. \tag{X-19}$$

The area element dS_0 originally perpendicular to n_i^0 ends up as the area element dS perpendicular to n_i; from Sec. I-12, the relation between the two is

$$n_i\, dS = J\left(\frac{x}{a}\right)\frac{\partial a_p}{\partial x_i}\, n_p^0\, dS_0,$$

where J is the Jacobian. Clearly

$$T_i^0\, dS_0 = T_i\, dS,$$

That is,

$$\tau_{ji}^0 n_j^0\, dS_0 = \tau_{it}(n_t\, dS)$$

$$= \tau_{it}\frac{\rho_0}{\rho}\frac{\partial a_j}{\partial x_t}\, n_j^0\, dS_0,$$

so that (Murnaghan)

$$\tau_{ji}^0 = \tau_{it}\frac{\rho_0}{\rho}\frac{\partial a_j}{\partial x_t}. \tag{X-20}$$

A consequence of Eq. (X-20) is that τ_{ij}^0 is not symmetric, a fact which would in any event be anticipated on physical grounds.

Define now F_i^0 as the current body force exerted on a unit mass originally at (a_i). Then, for any portion of material originally occupying a volume V_0 of surface S_0,

$$\int \rho_0 F_i^0\, dV_0 + \int T_i^0\, dS_0 = \int \rho_0 \ddot{u}_i\, dV_0,$$

from which, by use of Eq. (X-19) and the divergence theorem, it can be concluded that

$$\frac{\partial \tau_{ji}^0}{\partial a_j} + \rho_0 F_i^0 = \rho_0\frac{\partial^2 u_i}{\partial t^2}, \tag{X-21}$$

where the displacement u_i is expressed in terms of Lagrangian variables.

To derive a general stress-strain relation, consider the rate at which surface and body forces are doing work on the material originally occupying V_0, S_0:

$$\dot{W} = \int \rho_0 F_i^0 \dot{u}_i \, dV_0 + \int T_i^0 \dot{u}_i \, dS_0$$
$$= \dot{K} + \int \tau_{ji}^0 \frac{\partial \dot{u}_i}{\partial a_j} \, dV_0,$$

just as in Eq. (X–1). Again, $\dot{K}$ may be omitted for sufficiently slow motions; defining ϕ to be the free energy per unit mass, an isothermal process requires (cf. the derivation of Eq. (X–2)) that

$$\tau_{ji}^0 \frac{\partial \dot{u}_i}{\partial a_j} = \rho_0 \dot{\phi}$$
$$= \rho_0 \frac{\partial \phi}{\partial(\partial u_p/\partial a_q)} \frac{\partial \dot{u}_p}{\partial a_q},$$

where ϕ is considered to be a function of the η_{ij} and so of the $\partial u_i/\partial a_j$, and of T. Because this must hold for all possible motions, we obtain (Kirchhoff)

$$\tau_{ji}^0 = \rho_0 \frac{\partial \phi}{\partial(\partial u_i/\partial a_j)}. \tag{X–22}$$

The fact that ϕ is a function of the $\partial u_i/\partial a_j$ only via its functional form in terms of (η_{ij}) implies of course that ϕ is not a completely general function of the former. It is easily checked that Eqs. (X–3) and (X–22) are equivalent.

One other kind of stress is of interest. Consider a bundle of material fibers having a cross-sectional area dS_0 before straining. After the motion, the cross-sectional area of the fiber bundle is clearly

$$dS = \frac{\rho_0}{\rho} \frac{ds_0}{ds} \, dS_0,$$

where ds_0, ds are the initial and final lengths of one of the elemental fibers. In some problems, the total tensile force exerted by this bundle of fibers — that is, the normal stress on dS — is conveniently thought of as a stress which "follows the fibers" as the fibers move; let us develop an exact expression for its magnitude. If a generic fiber has original components da_i and final components dx_i, then the unit normal to the area dS is dx_i/ds, so that the normal force is

$$N = \tau_{ij} \frac{dx_i}{ds} \frac{dx_j}{ds} \, dS$$
$$= \left(\rho \frac{\partial \phi}{\partial \eta_{pq}} \frac{\partial x_i}{\partial a_p} \frac{\partial x_j}{\partial a_q}\right) \left(\frac{\partial x_i}{\partial a_r} \frac{da_r}{ds} \frac{\partial x_j}{\partial a_t} \frac{da_t}{ds}\right) dS.$$

Using

$$\frac{\partial x_i}{\partial a_p}\frac{\partial x_i}{\partial a_r} = 2\eta_{pr} + \delta_{pr},$$

and substituting for dS and for (ds_0/ds), the last expression becomes

$$N = \rho_0\, dS_0 \frac{da_r}{ds_0}\frac{da_t}{ds_0}\left(1 + 2\eta_{ij}\frac{da_i}{ds_0}\frac{da_j}{ds_0}\right)^{-\frac{3}{2}}\left(\frac{\partial\phi}{\partial\eta_{rt}} + 4\frac{\partial\phi}{\partial\eta_{pt}}\eta_{pr} + 4\frac{\partial\phi}{\partial\eta_{pq}}\eta_{pr}\eta_{qt}\right). \tag{X–23}$$

Index

Absolute temperature, 157
Acceleration, 66, 87
Adiabatic buckling, 207
Adiabatic bulk modulus, 161, 163, 164
Analogies, 90
Anisotropy, 107
Approximate variational methods, 140
Aquaro sphere theorem, 115

Beltrami-Michell equations, 86
Beltrami solution of wave equation, 181
Bessel inequality, 132
Betti's integrals, 118
Betti-Rayleigh reciprocal theorem, 103, 113
Betti representation, 121
Bianchi identity, 82
Biezeno, C. B., 205
Biharmonic function, 90, 116, 120, 126
Binormal, 17
Body force: concentrated, 92, 153, 186
  removal of, 91
  thermoelastic, 171
Bounds: on eigenvalues, 197
  on stress and displacement, 153
Bryan, G. H., 205
Bulk modulus, 84, 161
Busemann, A., 59

Calculus of variations, 137
Carnot, S., 156
Cartesian coordinates, 1
Cartesian tensors, 37, 39
Castigliano's theorem, 148
Cauchy conditions, 180
Cauchy molecular model, 109
Center of expansive pressure, 94, 153
Centers, line of, 95
Characteristics, 177
Clausius, R., 155, 156
Coefficient of thermal expansion, 161
Cofactor, 14; of direction cosines, 32
Cohn-Vossen, S., 58
Collatz, L., 197
Comma notation, 12
Compatibility equations, 78, 86
  curvilinear coordinates, 106
  large displacements, 81
  special form, 128
  in variational methods, 148, 149
Complementary energy, 150
Completeness of equation set, 100, 141
Concentrated body force, 91, 153, 171, 186
Concentrated impulsive pressure, 188
Concentrated moment, 94, 153
Concentrated moving force, 187
Concentrated shear, 153
Concentrated surface force, 95, 96
Continuity, equation of, 66
Contraction of tensor, 41
Coordinate system: attached to deformed body, 68
  Cartesian, 1
  curvilinear, 104
  linearizing, for strain, 73
  normal, 194
  principal, 54
  rotation, 30
  stress, 60
  on surface, 18
Coulson, C. A., 182
Courant, R., 143, 151, 194
Cross product, 6, 9
Curl, 13
  in curvilinear coordinates, 26, 103
Curvature: of space curve, 17
  of surface, 22

δ, symbol, 10
∇, symbol, 12, 13, 105
Dead loading, 205
Deformation: of body, 65
  of surface, 20
Delta function, 112, 183
Density after deformation, 77, 78
Depth of Rayleigh waves, 192
Determinants, 14
  direction cosine, 32
  functional, 15

Diaz, J. B., 116, 132, 153
Differential geometry: of space curve, 16
  of surface, 18
Differential identities, 159, 161
Differentiation behind integral, 67
Dilatation, Green's functions for, 119
Dirac delta function, 112, 183
Direction cosines, 31
Dirichlet-type problem, 88, 112, 114, 180
Discontinuity: displacement, 99
  in wave propagation, 181
  stress, 100
Dislocations, 100
Displacement: boundary condition, 114
  bounds, 153
  general solution for, 110
  large, 203
  relation to strain, 65
  special functions, 120
  sphere theorem, 115
Divergence: in curvilinear coordinates, 26, 103
  physical interpretation, 24
  theorem, 23
  of vector, 12
Double force without moment, 94
Dyad, 38
Dyadic circle, 63

e, symbol, 9
e–δ identity, 10, 45
Eason, G., 187
Eigenfunctions and eigenvalues, 51, 194
Elastic constants, relations between, 85, 109, 164
Elastic energy, 101, 108
Elastic modulus, 84, 165
Elastic stability, 205
Elastic range, 198
Elastic waves, 172
Elasticity: analogies, 90
  boundary value problems, 88
  curvilinear equations, 104
  linear, 83
  nonlinear, 198
  summarized equations, 86
  thermoelasticity, 155
  time-dependent, 172
  variational methods, 137
Elliptic point, 21
Energy: criterion in stability, 205
  and function space, 130, 151
  flux, 177
  free, 160, 198
  identity, 101
  internal, 159, 198
  minimal complementary, 150, 151
  minimal potential, 150, 152
  in nonlinear elasticity, 198
  power series expansion, 109, 200, 203
  rate of doing work, 108, 198
  stress and strain, 101
  and virtual stress, 147
  and virtual work, 137
Enthalpy, 160
Entropy, 157, 198
Equilibrium, equations of, 50, 86
  alternative equations, 87
  curvilinear coordinates, 106
  general solutions, 126, 129
Eubanks, R. A., 96, 99, 125, 188
Euler's rotation formula, 33
Eulerian rotation tensor, 74
Eulerian stress-strain law, 200
Eulerian strain tensor, 69
Eulerian variables, 65
Ewing, W., 194
Expansion coefficient, thermal, 161

Flux of energy, 177
Follow-the-fiber stress, 211
Fourier transforms, 185, 192
Free energy, 160, 198
Free enthalpy, 160
Frenet-Serret formulas, 18
Friedrichs, K. O., 143, 151
Fulton, J., 187
Function space, 130
  bounds on stress and strain, 153
  hypercircle, 134, 136
  inequalities, 135, 136
  orthonormal vectors, 132, 133
  and variational methods, 151
Functional transformation, 15
Fundamental solution: of elasticity equations, 91, 113
  of Laplace's equation, 112

Galerkin representation, 125
Galerkin variational procedure, 142
Gaussian curvature, 22
Gauss's theorem, 23
General elastic stability, 205

ɜeneral solutions: of elasticity equations, 114, 116
of Laplace's equation, 110
and thermoelasticity, 171
of wave equation, 176, 178, 181
ɜeneralized Hooke's law, 83
ɜoodier, J. N., 205
ɜradient, 12: in curvilinear coordinates, 105
in strain space, 203
of stress function, 60
ɜram-Schmidt procedure, 132
ɜreen's function: for elasticity equations, 114, 116, 118
for Laplace's equation, 112
for wave equation, 183, 188
ɜreen's identities, 25, 110, 119
ɜreenberg, H., 153, 204

Iarmonic functions, 110
solutions in terms of, 120
Ieat: conduction, 168
generated by compression, 163
and thermodynamics, 155
Ielmholtz's equation, 185
Ielmholtz's theorem, 27, 129
Iencky, H., 205
Iilbert, D., 58, 151, 194
Iooke's law, 83
Iuygen's principle, 182
Iyperbolic point, 21
Iypercircle, in function space, 134, 136

ncompressibility, 90
ndex notation, 3, 12
nequalities: Bessel, 132
in function space, 135, 136
Schwartz, 131
on stress and strain, 153
triangle, 131
nfinitesimal rotations, 32
order of, 36
nfluence coefficients, 103
nfluence triangle, 179
nner product, 5
ntegral equation, 197
nvariants: of stress, 56
of strain, 77, 203
rrotational vector field, 26
rrotational wave, 173
sothermal bulk modulus, 161, 163, 164
sotropy: of elastic body, 84
in nonlinear elasticity, 200
of tensor, 43, 44

Jacobian, 15
Jacovache, L., 188
Jardetzky, W., 194

Kellogg, O. D., 93
Kelvin, Lord (W. Thomson), 91, 121, 122, 157
Kepler's law, 9
Kinetic energy, 108, 177, 198
Kirchhoff, G., 182, 211
Kron, G., 91

Lagrangian multipliers, 141, 144, 146, 148
Lagrangian rotation tensor, 73
Lagrangian strain, 69
Lagrangian variables, 65
Lamb, H., 188
Lamé's constants, 84
Laplacian, 13
Green's functions, 112, 183
physical interpretation, 25
Linear algebraic equations, 14, 51
Linear elasticity, 84
Love, A.E.H., 86, 109, 186

Magnitude of vector in function space, 131
Material derivative, 66
in curvilinear coordinates, 106
in travelling wave, 173
Maxwell's equations (thermodynamics), 160, 165
Maxwell-Morera representation, 126
Mean curvature, 22
Minimal complementary energy, 150, 151
Minimal potential energy, 150, 152
Mixed boundary-value problem, 118, 120
Mixed thermoelastic equations, 168
Mohr circle, 61
Moving concentrated force, 187
Multiple-valuedness, 27
Murnaghan, F., 199, 200, 210

Natural boundary conditions, 139, 143
Navier's equations, 86
Neuber, H., 124, 188, 205
Neumann-type problem, 88, 112, 116, 180
Nominal stress, 210
Nonlinear elasticity, 198

Nonlinear adiabatic and isothermal motion, 199
Nonlinear stress-strain law, 199, 200, 202, 211
Nonlinearity, partial, 203
Normal component of stress, 50
Normal curvature, 22
Normal mode, 194
Normal vector: to area element, 46
  to surface, 19
Novozhilov, V. V., 74, 205

Orthogonality: in function space, 132
  of normal modes, 196
  of principal axes, 53
Orthonormal vectors, 132
  complete set, 134
Osculating plane, 17

Papkovich-Neuber representation, 124, 188
Parabolic point, 21
Partial nonlinearity, 203
Payne, L. E., 116
Phase difference in Rayleigh waves, 192
Plass, H. J., 205
Poisson equation, 111, 118
Poisson formula for wave equation, 183
Poisson's ratio, 84
  bounds on, 85
  negative, 86
Polarization of elastic waves, 193
Potential, retarded, 182
Potential energy, 150, 177
Potential function, 27
Potential theory, 110
Prager, W., 132, 205
Press, F., 194
Pressure loading, 205, 209
Principal axes, 51, 53, 76
Principal axis coincidence, 86, 202
Principal directions, 22, 54
Principal normal curvature, 22
Principal normal vector, 16
Principal shears, 62
Principle: of virtual stress, 147
  of virtual work, 137
Projection theorems, 56
Propagation of signal, 174
Pythagoras' theorem, generalized, 132, 135

Radiation condition, 185
Rayleigh quotient, 197
Rayleigh-Ritz methods, 140, 142, 147, 151, 197
Rayleigh waves, 188, 190
Reciprocal theorem of Betti and Rayleigh, 103, 113, 171
Reciprocal relation of energy theorems, 151
Reflection of elastic waves, 193
Refraction of elastic waves, 194
Reissner, E., 149
Resolution of wave equation, 176
Retarded potential, 182
Reversibility (thermodynamics), 156, 198
Rigid-body motion, 70, 117
Rigidity theorem, 103
Roderigues' rotation formula, 35
Rotation: bounds, 153
  of coordinate system, 30
  of deformed body, 73, 74
  Green's functions for, 120
  harmonicity, 90
  identity, 87
  of rigid body, 33

Scalar product, 4, 131
Schild, A., 82
Schwartz inequality, 131
Shear, pure, 57, 167
Shear stress, 50, 58
Shear deformation, 71
Signal-propagation speed, 175
Singularities, 91, 94, 96, 97, 99
Sneddon, I., 187
Snell's law, 194
Sokolnikoff, I. S., 87
Solenoidal vectors, 27
Solenoidal waves, 173
Somigliana integrals, 113
Soroka, W. W., 91
Southwell, R. V., 295
Space curves: canonical representation, 1
  differential geometry, 16
  motion of orthogonal triad, 36
Special functions: solutions in terms c 120, 188
  reduction in number of, 125
Specific heats, 161, 165
Sphere theorems, 115, 116, 119
St. Venant's principle, 90

Stability criterion, 207, 208, 209
in curvilinear coordinates, 208
Stability, elastic, 205
Star-shaped region, 122, 125
Steel, thermoelastic properties, 162, 166, 167
Sternberg, E., 96, 99, 125, 188
Stokes, G. G., 187
Stoke's theorem, 25
Stoke's corollary, 26
Strain: bounds, 153
compatibility, 78, 81
correction to linear approximation, 75
decomposition, 129
derivatives, 72
energy, 101
function space, 130
geometry, 77
greatest and least, 76
invariants, 77, 78
linearizing coordinate system, 73
nonlinearity, 71, 72, 203
physical interpretation, 70, 76
principal axes, 76
projection theorems, 77
pure shear, 77
symmetry, 70
tensor, 68
vector, 75, 130
Stream function, 27
Stress: and specific heats, 166
bounds, 153
decomposition, 129
deviator, 59
energy, 101
function space, 130
geometry, 58
invariants, 56
maximal normal, 55
Mohr circle, 61
on moving area element, 210
nominal, 210
in nonlinear elasticity, 198
normal and shear components, 50
principal axes, 51
projection theorems, 56
pure shear, 57
quadratic, 60
special functions, 126
symmetry of tensor, 51
tensor, 48, 49
vector, 46, 130
virtual, 147
Stress boundary conditions, 116
Stress-strain laws: in linear elasticity, 83, 107
in nonlinear elasticity, 19, 20, 203, 211
Subscript notation, 3, 12
Substitution operator, 10
Summation convention, 6
Superposition, 87
of normal modes, 195
Surface-volume integrals, 23
Surfaces: differential geometry, 18
stress quadric, 60
Surface waves, 190
Synge, J. L., 82, 115, 116, 132, 153

Temperature, thermodynamic, 157
Tensor: decomposition, 129
definition and properties, 37
equations, 41
isotropic, 43, 44
order of, 39
strain, 68, 106
stress, 48, 106
symmetry, 42
tests, 43
Test for tensor character, 43
Thermodynamics: Carnot's theorem, 156
differential identities, 159, 161, 166
enthalpy, 160
entropy, 157, 198
free energy, 160, 198
free enthalpy, 160
isothermal and adiabatic moduli, 160, 163, 164
laws of, 155
Maxwell's equations, 160, 165
in nonlinear elasticity, 198
reversibility, 156
of simple substance, 159
specific heats, 160, 165
temperature, 157
Thermoelasticity, 155, 169
Torque vector, 8
Torsion of space curve, 17
Total curvature, 22
Trefftz, E., 205
Triangle inequality, 131

Undetermined constants, method of, 141
Uniqueness: and completeness, 100

Uniqueness (*cont.*)
and dislocations, 100
in nonlinear elasticity, 205
and singularities, 99
Uniqueness theorem, 97
Unit vectors in function space, 132, 133

Variational methods, 137
approximate methods, 140, 143
Lagrangian multipliers, 141, 144
natural boundary conditions, 137
in nonlinear elasticity, 204
Rayleigh quotient, 197
reciprocal principle, 143
reciprocal relations, 151
Reissner method, 149
subsidiary conditions, 143
and thermoelasticity, 171
use of, 139
virtual stress, 147
virtual work, 137
Vectors:
algebra, 2
Bessel inequality, 132
binormal, 17
comma notation, 12
curl, 13, 105
curvilinear coordinates, 105
differentiation, 8, 12, 13, 105
divergence, 12, 105
in function space, 130
gradient, 12, 105
index notation, 3, 12
outward normal, 26
principal normal, 16
scalar product, 4, 105, 131
Schwartz inequality, 131
strain, 75
stress, 46
summation convention, 6
unit tangent, 16
vector product, 6, 9, 105
Velocity, 66
Vibrations, 194
Virtual stress, 147
accuracy, 142
approximate methods, 148
and complementary energy, 150
and Lagrangian multipliers, 148
converse, 147
in nonlinear elasticity, 204
Virtual work, 137
accuracy, 142
approximate methods, 140
and Lagrangian multipliers, 146
and potential energy, 150
converse theorem, 138
in nonlinear elasticity, 204
Voigt, W., 109
Volterra, E., 100
Volume of parallelepiped, 8, 145

Washizu, K., 153
Wave, elastic, 172
boundary conditions, 174, 179
characteristics, 177
energy flux, 177
equation, 173
general solutions, 181, 185
Green's functions, 183
longitudinal, 173
Rayleigh, 188, 190
reflection, 193
resolution of, 177
standing, 194
trains, 183
transverse, 173
Weingarten, V. I., 100

Youngs' modulus, 84

Ziegler, H., 205